T0290922

IET ENERGY ENGINEERING SERIES 210

Superconducting Magnetic Energy Storage in Power Grids

Other volumes in this series:

Superconducting Magnetic Energy Storage in Power Grids

Edited by
Mohd. Hasan Ali

The Institution of Engineering and Technology

Published by The Institution of Engineering and Technology, London, United Kingdom

The Institution of Engineering and Technology is registered as a Charity in England & Wales (no. 211014) and Scotland (no. SC038698).

The Institution of Engineering and Technology
Futures Place
Kings Way, Stevenage
Herts, SG1 2UA, United Kingdom

www.theiet.org

British Library Cataloguing in Publication Data
A catalogue record for this product is available from the British Library

ISBN 978-1-83953-500-0 (hardback)
ISBN 978-1-83953-501-7 (PDF)

Typeset in India by Exeter Premedia Services Private Limited
Printed in the UK by CPI Group (UK) Ltd, Croydon
Cover Image: Mark Garlick/Science Photo Library via Getty Images

Contents

About the Editor

Dr. Mohd Hasan Ali received his Ph.D. degree in electrical and electronic engineering from Kitami Institute of Technology, Kitami, Japan, in 2004. He is currently an Associate Professor with the Electrical and Computer Engineering Department at the University of Memphis, USA, where he leads the Electric Power and Energy Systems (EPES) Laboratory. He held research positions in Canada, Japan, and South Korea. His research interests include smart-grid and microgrid systems, cybersecurity issues and solutions to modern power grids, electric vehicle charging system and station, renewable energy systems, energy storage systems, and load forecasting in smart buildings. Dr. Ali has more than 210 publications including 4 books, 6 book chapters, 3 patents, 71 top ranked journal papers, 100 peer-reviewed international conference papers and 20 national conference papers. According to Google Scholar, as of November 2022, the total citations number of his published research is 5,152 with an h-index of 36 and i10-index 103. His research has been funded by various sponsors such as the National Security Agency (NSA), Department of Energy (DOE), National Science Foundation (NSF), the American Public Power Association (APPA), the Idaho National Laboratory (INL), Electric Power Research Institute (EPRI), to name a few. He received the University of Memphis Alumni Association Distinguished Teaching Award in 2020 and the Outstanding Faculty Teaching Award from the Herff College of Engineering in 2019. He serves as the Editors of the IEEE Transactions on Sustainable Energy, IEEE Transactions on Energy Conversion, IEEE Power Engineering Letters, IET-Generation, Transmission and Distribution (GTD) journal, MDPI Electronics Journal, and Frontiers in Energy Research. Dr. Ali is a Senior Member of the IEEE Power and Energy Society (PES). Also, he is the Chair of the PES of the IEEE Memphis Section.

Chapter 1

Introduction

Mohd Hasan Ali[1]

1.1 Need to defossilize or decarbonize

Decarbonization is the term used for removal or reduction of carbon dioxide (CO_2) output into the atmosphere. Decarbonization has become a global imperative and a priority for governments, companies, and society at large, because it plays a very important role in limiting global warming. Many companies across all industries (e.g., in energy, transport, and consumer products) have publicly declared their intention to become carbon neutral by 2050 [1]. Carbon neutral (i.e., net zero) means that all greenhouse gas emissions produced are counterbalanced by an equal amount of emissions that are eliminated. Achieving this will require rapid decarbonization [2].

Ambitious attempts to avoid a rise higher than 1.5 °C in the global average temperature require a rapid reduction in the utilization of fossil fuels across all energy sectors [3]. The decreasing cost of wind and solar power has made this a more attractive economic prospect. Electrification as a means to decarbonize many services across all energy sectors adds to the need to understand and develop 100% renewable energy systems [3].

There are two aspects to decarbonization. The first entails reducing the greenhouse gas emissions produced by the combustion of fossil fuels. This can be done by preventing emissions through the use of zero-carbon renewable energy sources such as wind, solar, hydropower, geothermal, and biomass, which now make up one-third of global power capacity and electrifying as many sectors as possible. Energy efficiency will reduce the demand for energy, but increasing electrification will increase it, and in 2050, the demand for power is expected to be more than double what it is today [2].

Furthermore, to achieve decarbonization, all aspects of the economy must change – from how energy is generated, and how we produce and deliver goods and services, to how lands are managed. The CO_2 and methane emissions that are warming the planet come largely from the power generation, industry, transport,

[1]The University of Memphis, Memphis, Tennessee, United States

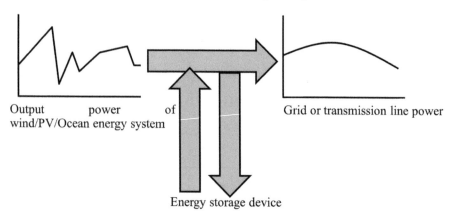

Output   power   of
wind/PV/Ocean energy system

Grid or transmission line power

Energy storage device

Figure 1.1   Role of energy storage device.

buildings, and agriculture and land-use sectors of the global economy, so these sectors must all be transformed.

1.2   Need for storage system to integrate renewables

For nearly a century, global power systems have focused on three key functions: to generate, transmit, and distribute electricity as a real-time commodity. Physics requires that electricity generation always be in real-time balance with load, despite variability in load on timescales ranging from subsecond disturbances to multiyear trends. The increasing use of renewable energy sources during the last two decades has increased the importance of research and development of energy storage systems. Intermittent sources such as wind, solar, or tide do not always generate energy at the same rate as the energy in cities is consumed. This switch, from energy systems governed by traditional fossil fuels to systems with high penetration of renewable energies, introduces load imbalances between supply and demand. Thus, grid operators are using new methods to maintain this balance. Energy storage, if suitably deployed, gives system operators a flexible and fast response resource to effectively manage variability in generation and load [4].

Figure 1.1 shows the role of an energy storage system in a smart grid/microgrid. According to system power requirements, energy storage device should be capable of charging and discharging quickly. The possible benefits of energy storage devices include transmission enhancement, power oscillation damping, dynamic voltage stability, tie line control, short-term spinning reserve, load leveling, subsynchronous resonance damping, power quality improvement, etc.

Some of the well-known energy storage systems are mentioned as follows: (i) superconducting magnetic energy storage (SMES), (ii) battery energy storage (BES), (iii) ultra/supercapacitor energy storage (SCES), (iv) flywheel energy storage (FES), (v) pumped hydro energy storage (PHES), and (vi) thermal energy storage (TES).

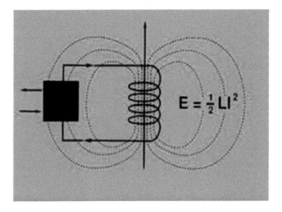

Figure 1.2   SMES working principle

1.2.1   Superconducting magnetic energy storage

SMES is a large superconducting coil capable of storing electric energy in the magnetic field generated by DC current flowing through it. The real power as well as the reactive power can be absorbed by or released from the SMES coil according to system power requirements. Although superconductivity was discovered in 1911, SMES has been under study for electric utility energy storage application since the early 1970s [5]. SMES systems have attracted the attention of both the electric utilities and the military due to some salient features such as fast response (millisecond/megajoule), high efficiency (a charge–discharge efficiency over 95%), and ability to control both active and reactive powers. Possible applications of SMES include load leveling, dynamic stability, transient stability, voltage stability, frequency regulation, transmission capability enhancement, power quality improvement, automatic generation control, uninterruptible power supplies, etc. The one major advantage of the SMES coil is that it can discharge large amounts of power for a small period of time. Also, unlimited number of charging and discharging cycles can be carried out.

Figure 1.2 shows an SMES working principle. For an SMES system, the inductively stored energy (E in Joule) and the rated power (P in Watt) can be expressed as follows:

$$E = \frac{1}{2}LI^2 \tag{1.1}$$

$$P = \frac{dE}{dt} = LI\frac{dI}{dt} = VI \tag{1.2}$$

where L is the inductance of the coil, I is the DC current flowing through the coil, and V is the voltage across the coil.

An SMES unit consists of a large superconducting coil at the cryogenic temperature. This temperature is maintained by a cryostat or dewar that contains helium or nitrogen liquid vessels. A bypass switch is used to reduce energy losses when

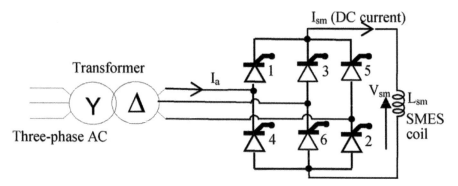

Figure 1.3   SMES unit with six-pulse bridge AC/DC thyristor-controlled converter [5]

the coil is on standby. And it also serves other purposes such as bypassing DC coil current if utility tie is lost, removing converter from service, or protecting the coil if cooling is lost [5].

In SMES systems, it is the power conditioning system (PCS) that handles the power transfer between the superconducting coil and the AC system. According to topology configuration, there are three kinds of PCSs for SMES, namely, the thyristor-based PCS, voltage source converter (VSC)-based PCS, and current source converter (CSC)-based PCS.

Figure 1.3 shows the basic configuration of a thyristor-based SMES unit, which consists of a wye-delta transformer, an AC–DC thyristor-controlled bridge converter, and a superconducting coil or inductor. Figure 1.4 shows the basic configuration of the VSC-based SMES unit, which consists of a wye-delta transformer, a six-pulse pulse width modulation (PWM) rectifier/inverter using insulated gate

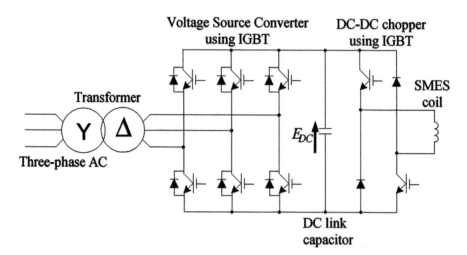

Figure 1.4   Basic configuration of VSC-based SMES system [5]

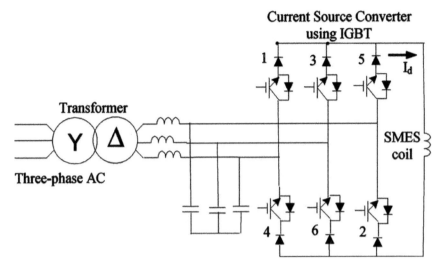

Figure 1.5   SMES system with a CSC [5]

bipolar transistor (IGBT), a two-quadrant DC–DC chopper using IGBT, and a super-conducting coil or inductor. The PWM converter and the DC–DC chopper are linked by a DC link capacitor. Figure 1.5 shows the basic configuration of the CSC-based SMES unit. The DC side of CSC is directly connected with the superconducting coil, and its AC side is connected to the power line. A bank of capacitors connected to a CSC input terminal is utilized to buffer the energy stored in line inductances in the process of commutating direction of AC line current. Furthermore, the capacitors can filter the high-order harmonics of the AC line current.

1.2.2   Battery energy storage

Batteries are one of the most cost-effective energy storage technologies available, with energy stored electrochemically. A battery system is made up of a set of low-voltage/power battery modules connected in parallel and series to achieve a desired electrical characteristic. Batteries are "charged" when they undergo an internal chemical reaction under a potential applied to the terminals. They deliver the absorbed energy, or "discharge," when they reverse the chemical reaction. Key factors of batteries for storage applications include: high energy density, high energy capability, round trip efficiency, cycling capability, and initial cost.

Lithium-ion, lead-carbon, sodium-sulfur, and redox-flow batteries are the main battery technologies that are used for energy storage. With the continuous release of support policies and improvements of the manufacturing processes, the BES technology has developed rapidly. Its key technical indicators, such as battery safety, cycle life, and energy density, have greatly improved. On the other hand, the application costs have dropped significantly [6].

1.2.3   Ultra/supercapacitor energy storage

In recent years, due to its cost-effectiveness, the SCES system has been extensively used for dynamic performance enhancement of power grids [7, 8]. The SCES can control both active and reactive powers quickly and simultaneously. A supercapacitor (SC), also known as double-layer capacitor, differs from a regular capacitor in that it has very high capacitance. An SC is constructed with two porous activated carbon electrodes impregnated with electrolyte and separated by a porous insulating membrane. When a voltage is applied to the SC terminals, a double layer is formed at the interface between the electrode and the electrolyte. The energy storage mechanism is primarily electrostatic rather than faradaic [9]. The cell voltage of SC is low (typical value is 2.7 V), and thus, several cell units should be connected in series to form the rated voltage in an SC bank. The SC is widely used as energy storage device in industry, electric vehicles, and renewable energy applications (solar farms and wind turbines) individually or in combination with batteries [9].

1.2.4   Flywheel energy storage

Flywheels can be used to store energy for power systems when the flywheel is coupled to an electric machine [10, 11]. In most cases, a power converter is used to drive the electric machine to provide a wider operating range. Stored energy depends on the moment of inertia of the rotor and the square of the rotational velocity of the flywheel, as shown in (1.3).

$$E = \frac{1}{2}I\omega^2 \tag{1.3}$$

$$I = \frac{r^2 mh}{2} \tag{1.4}$$

The moment of inertia (I) depends on the radius, mass, and height (length) of the rotor, as shown in (1.4). Energy is transferred to the flywheel when the machine operates as a motor (the flywheel accelerates), charging the energy storage device. The flywheel is discharged when the electric machine regenerates through the drive (slowing the flywheel).

The energy storage capability of flywheels can be improved either by increasing the moment of inertia of the flywheel or by turning it at higher rotational velocities, or both. Figure 1.6 shows an application of FES system coupled to a dynamic voltage. If an FES system is included with a flexible AC transmission system device and a DC bus, an inverter is added to couple the flywheel motor/generator to the DC bus. FES has been considered for several power system applications, including power quality applications as well as peak shaving and stability enhancement.

1.2.5   Pumped hydro energy storage

Hydroelectric storage is a process that converts electrical energy to potential energy by pumping water to a higher elevation, where it can be stored indefinitely and then released to pass through hydraulic turbines and generate electrical energy. A PHES can respond quickly to mismatches between demand and generation [12]. A typical

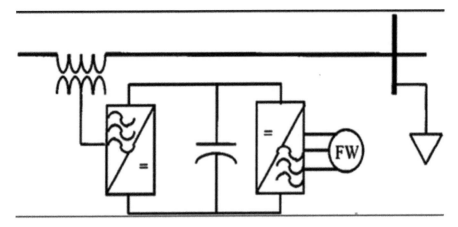

Figure 1.6   FES coupled to a dynamic voltage

pumped storage development is composed of two reservoirs of equal volume situated to maximize the difference in their levels. These reservoirs are connected by a system of waterways along which a pumping-generating station is located. The principal equipment of the station is the pumping-generating unit, which is generally reversible and used for both pumping and generating, functioning as a motor and pump in one direction of rotation and as a turbine and generator in opposite rotation.

1.2.6   Thermal energy storage

TES is one form of energy storage. In this case, a material gains energy when increasing its temperature and loses it when decreasing. Taking advantage of this property makes it possible to use different materials with different thermal properties and achieve various results which can lead to different TES applications (e.g., heating and cooling). TES can help balance energy demand and supply on a daily, weekly, and even seasonal basis, presented in thermal systems. TES can also reduce peak demand, energy consumption, CO_2 emissions, and costs while also increasing the overall efficiency of energy systems [13].

The most common application for TES is in solar thermal systems. However, due to its wide range of benefits, TES is used in many other applications as well – such as those found in CELSIUS demonstrators to store heat in building structures, to couple waste heat and district heating systems, and to couple heat pumps and combined heat and power generators in district heating networks [13].

1.2.7   Hybrid energy storage devices

Certain applications might require a combination of energy, power density, cost, and life cycle specifications that cannot be met by a single energy storage device. To implement such applications, hybrid energy storage devices (HESDs) have been proposed. HESDs electronically combine the power output of two or more devices with complementary characteristics. HESDs all share a common trait, combining high power devices (devices with quick response) and high energy devices (devices with slow response). Some HESDs are listed below: (i) battery and SCES; (ii) fuel cell and SCES; (iii) battery and flywheel; and (iv) battery and SMES.

1.3   Shortcomings of other storage technologies

Although the major problems of confronting the implementation of SMES units are the high cost and environmental issues associated with strong magnetic field, there are disadvantages of other energy storage systems, as noted below.

- Some of the disadvantages of battery energy storage system (BESS) include limited life cycle, voltage and current limitations, and potential environmental hazards.
- Some of the disadvantages of pumped hydroelectric are large unit sizes, and topographic and environmental limitations.
- Relatively short duration, high frictional loss (windage), and low energy density restrain the flywheel systems from the application in energy management.
- Similar to flywheels, the major problems associated with SC are the short durations and high energy dissipations due to self-discharge loss.
- TES has several disadvantages [14]. (i) It is considerably polluted by the continuous emissions of greenhouse gases. (ii) The vapor and heat emissions generated in the thermal plants negatively affect the climate of the place where they are located. (iii) The water used during the process ends up contaminated. (iv) By using nuclear technology, it leaves a lot of radioactive waste. (v) By using fossil fuels, the long-term generation of this energy depends on the fossil reserves available. (vi) River ecosystems are severely affected, as they are spaces where the hot water released from the plants is poured out. (vii) Releases contaminants with CO_2. (viii) The spaces where the energy is captured are not affected as they do not decrease the amount of air in the place. (ix) The construction of the thermal power plants takes years. And (x) the transfer of energy is complicated.
- However, among all energy storage systems, SMES is the most effective from the viewpoints of fast response, charge and discharge cycles, control ability of both active and reactive powers simultaneously, etc.

1.4   Overview of the contents

The need to increase the share of renewable energy in the electric power grid brings with it the requirement to buffer its intermittency and to sustain high power quality. Energy storage is key to this requirement. While a number of storage technologies exist, each with advantages and disadvantages at the various voltage and grid levels, SMES can be playing an important role in the decarbonized energy system. Thus, research efforts are going toward developing more prototypes, gaining experience, and reducing cost, and widening the application scope of SMES. This book provides researchers as well as utility experts a concise overview of the technology, applications, and challenges. The book is organized as follows.

Chapter 2 deals with an in-depth overview of SMES technology. It discusses the basics of SMES including its types, dynamic models, material, challenges, cooling system, applications to power and energy systems, practical aspects, and cost analysis.

SMES systems exchange power with the distribution grids via a power electronic converter. Therefore, fast and robust control strategies are needed to guarantee the efficient performance of these converters. Chapter 3 explains about various control methods of SMES. VSC- and CSC-based control approaches of SMES are discussed in this chapter.

Transient stability issues in power grids have been an important subject matter for a long time. Chapter 4 deals with the transient stability enhancement of power grids by SMES. Transient stability improvement in both conventional source-based grids and renewable energy source-based grids with case studies is discussed in this chapter.

Enhancement of load frequency control in interconnected microgrids by SMES is discussed in Chapter 5. Minimization of frequency and tie-line power oscillations in both conventional source-based and renewable energy source-based interconnected microgrids is elaborated in this chapter. Control methods of SMES for load frequency control with case studies are explained as well.

Dynamic performance enhancement of power grids by co-ordinated operation of SMES and other control systems is discussed in Chapter 6. Various combinations such as SMES and BES, SMES and fuel cell, SMES and static var compensator, SMES and static synchronous compensator, SMES and fault current limiter, etc., have been considered for the analysis of dynamic performance enhancement of power grids.

Chapter 7 deals with artificial intelligence (AI)-based controls of SMES for dynamic performance enhancement of power grids. Various AI-based methods such as fuzzy logic, neural network, adaptive neuro-fuzzy inference system, etc., have been used for SMES control in power grids.

Cybersecurity issues in power grids have been a hot topic in recent years. Therefore, cybersecurity problems in intelligent control-based SMES system and associated mitigation strategies are discussed in Chapter 8. Explanation has been provided on how cybersecurity issues happen in SMES system. Also, impacts of cyberattacks on SMES control system with case studies are discussed.

Finally, Chapter 9 provides an outlook including summary of findings, open challenges, and unsolved problems of SMES.

References

[1] [online]. Available from https://www2.deloitte.com/nl/nl/pages/energy-resources-industrials/articles/what-is-decarbonisation.html

[2]   [online]. Available from https://news.climate.columbia.edu/2022/04/22/what-is-decarbonization-and-how-do-we-make-it-happen/

[3]   Holttinen H., Kiviluoma J., Flynn D., *et al.* 'System impact studies for near 100 % renewable energy systems dominated by inverter based variable generation'. *IEEE Transactions on Power Systems.* 2022, vol. 37(4), pp. 3249–58.

[4]   Stenclik D., Denholm P., Chalamala B.R. *The role of energy storage for renewable integration* [online]. Available from https://www.osti.gov/servlets/purl/1429811

[5]   Ali Mohd.H., Wu B., Dougal R.A. 'An overview of SMES applications in power and energy systems'. *IEEE Transactions on Sustainable Energy.* 2010, vol. 1(1), pp. 38–47.

[6]   Li X., Wang S., China Electric Power Research Institute. 'A review on energy management, operation control and application methods for grid battery energy storage systems'. *CSEE Journal of Power and Energy Systems.* 2019, vol. 7(5), pp. 1026–40.

[7]   Hossain Md.R., Hossain Md.K., Ali Mohd.H., Luo Y., Hovsapian R. 'Synchronous generator stabilization by thyristor controlled supercapacitor energy storage system'. *SoutheastCon 2017*; Concord, NC, 2017.

[8]   Anwar A., Ali Mohd.H., Dougal R.A. 'Supercapacitor energy storage for low-voltage ride through in a 13.8KV AC system'. *Proceedings of the IEEE Southeast Con 2010*; North Carolina, 2010. pp. 189–92.

[9]   Yang H., Zhang Y. 'Analysis of supercapacitor energy loss for power management in environmentally powered wireless sensor nodes'. *IEEE Transactions on Power Electronics.* 2013, vol. 28(11), pp. 5391–403.

[10]   Tziovani L., Hadjidemetriou L., Charalampous C., Tziakouri M., Timotheou S., Kyriakides E. 'Energy management and control of a flywheel storage system for peak shaving applications'. *IEEE Transactions on Smart Grid.* 2021, vol. 12(5), pp. 4195–207.

[11]   Ghosh S., Kamalasadan S. 'An energy function-based optimal control strategy for output stabilization of integrated DFIG-flywheel energy storage system'. *IEEE Transactions on Smart Grid.* 2017, vol. 8(4), pp. 1922–31.

[12]   Bruninx K., Dvorkin Y., Delarue E., Pandzic H., Dhaeseleer W., Kirschen D.S. 'Coupling pumped hydro energy storage with unit commitment'. *IEEE Transactions on Sustainable Energy.* 2016, vol. 7(2), pp. 786–96.

[13]   [online]. Available from https://celsiuscity.eu/thermal-energy-storage/

[14]   [online]. Available from https://ventajasydesventajas.pro/en/energy/benefits-of-thermal-energy/

Chapter 2
Overview of SMES technology
Dumitru Cazacu[1] and Radu Jubleanu[2]

2.1 Introduction

The accelerated growth of energy consumption in recent times has led to the implementation of new production systems based on renewable sources (mainly solar and wind). Energy sources like fossil fuels can be used to provide energy according to customer demand, i.e., they are readily storable when not required. In contrast, renewable sources have a major disadvantage, namely, power fluctuations caused by primary sources (e.g., wind speed is not constant), which required the use of energy storage to ensure the balance between consumption and energy production. Energy storage is very important for electricity as it improves the way electricity is generated, delivered, and consumed. Energy storage technologies allow us to store excess energy and discharge it when there is low production or higher demand. They provide flexibility at different timescales – seconds/minutes, hours, weeks, and even months. Storage can help consumers increase self-consumption of solar electricity or to generate energy by providing flexibility to the system. The main storage technologies are batteries, flywheels, ultracapacitors, and superconducting magnetic energy storage (SMES) systems. Recent developments and advances in energy storage and power electronics technology are making the application of energy storage technologies a suitable solution for modern power applications. But other sources such as solar and wind energy need to be harvested when available and stored until needed. In this chapter, we want to approach different topologies of superconducting magnetic energy storage systems (EESs), which have as a central device the superconducting coil. This is considered the heart of superconductive devices, which must fulfill requirements such as low stray field and mechanical design suitable to support large Lorentz forces. However, superconducting coils cannot function without other essential components such as cooling devices, control and discharge cycle control equipment, temperature control devices, etc. All of these will be discussed in detail below. Different energy storage technologies are presented as well as the operating principle and the main components. Factors that influence the amount of

[1]University of Pitesti, Pitesti, Romania
[2]University POLITEHNICA of Bucharest, Romania

the stored energy, advantages, and drawbacks of this system by comparison with other technologies are included.

2.2   What is SMES?

2.2.1   SMES concept

The SMES is one of the very few direct electric EESs. Its energy density is limited by mechanical considerations to a rather low value on the order of 10 kJ/kg, but its power density can be extremely high. This makes SMES particularly interesting for high-power and short-time applications (pulse power sources). The SMES releases its energy very quickly and with an excellent efficiency of energy transfer conversion (>95%). Using SMES in energy storage can provide several advantages for energy systems, such as permitting increased penetration of renewable energy and better economic performance. The other technologies applied for utility applications are pumped hydroelectric systems, compressed air energy storage, and flow batteries. Many of these technologies have been initially used for large-scale load leveling applications. But now the EES is used to enhance the system stability, aid power transfer and improve power quality in power systems, frequency regulation, and damping energy oscillations [1]. The applications, advantages, and limitations of storage technologies in a power grid and transportation system are presented in Reference 2. Studied cases, state of storage technologies, and energy storage applications are depicted in Reference 3. A review regarding the energy storage types focused on operating principles and technological factors is included in Reference 4. In addition, a critical analysis of the various energy storage types is provided by reviewing and comparing the applications and technical and economic specifications of energy storage technologies. The energy storage application of green nanocomposites is described in Reference 5. The paper is concerned on the design and development of multifunctional energy storage materials.

High energy density SMES systems with second-generation (2G) high-temperature superconducting (HTS) is presented in Reference 6.

SMES is a direct energy storage device that stores energy in the form of direct current (DC) electricity that is the source of DC magnetic field. As SMES stores electrical current, the only conversion involved with the process is the conversion from alternating current (AC) to DC. In the last decades, SMES technology has become a promising high power storage research areas of superconductor applications, especially after the discovering of HTS materials in 1986. With significant progress in the manufacturing of 2G HTS tape, applications such as SMES have become promising for implementation in the electricity grid. The methodology for designing a 2G HTS SMES, using yttrium barium copper oxide (YBCO) tapes operating at 22 K is the main concern in Reference 7. Rong and Barnes [8] review the current status of HTS-based SMES technology as a developmental effort.

Energy storage units are used with renewable sources to act as backup units and store power when generated power is greater than the load power and discharge their power to the system in peak periods of load demand. This action reduces levels of fluctuations and acceptable limits of the system's frequency. In Reference 9, a robust control methodology for a wind–diesel–SMES system is described. Two controllers were designed to enhance the system performance.

2.2.2 The operating principle

An SMES consists of a superconducting coil maintained at temperatures below the critical value T_c, in order to maintain its superconductive state. The energy is stored in the magnetic field created by the current circulating in the superconducting coil. At those temperatures, the electrical resistance of the superconductors drops to almost zero, enabling the magnet to carry high currents without ohmic losses. High currents generate high value magnetic fields and this increases the amount of the stored magnetic energy. If a voltage is applied to the coil, the current in the coil increases as in (2.1).

$$U = L \frac{dI}{dt} \tag{2.1}$$

A magnetic field is generated and a certain magnetic energy is stored in the superconducting magnet. When an opposite voltage is applied to the coil, the current is reduced, i.e., the energy is extracted from the coil [5]. If the coil is perfectly short circuited by a superconducting connection, the current is kept indefinitely constant and the magnetic energy is stored. The resistance of the coil can be nonzero because of the current leads or of the connections between the coil elements. In this case, the time constant of the SMES is not infinite but has to be sufficiently long for the dedicated purpose [5]. The stored energy(w_m) can be expressed in different ways: by integration all over the space, by integration over the coil volume, or by using the self-inductance (L) of the coil and its current (I).

$$W_m = \int \frac{\overline{B}.\overline{H}}{2} dv = \int \frac{B^2}{2\mu_0.\mu_r} dv = \int \frac{\overline{A}.\overline{J}^2}{2} dv \tag{2.2}$$

The last expression is similar to that of the energy stored in a capacitor. SMES and capacitors supply the power grid without energy conversion. An SMES can be understood as an ideal current source by comparison with capacitors that are ideal voltage sources [10].

For the first two expressions, the integration is performed on the whole volume of the magnetic field. As it can be noticed from the second relation in (2.2), the energy should be stored, where μ_r is minimal, this means in the air or vacuum. Using magnetic material does, therefore, not improve the storage capacities and may only be used for magnetic flux density concentration. If magnetic materials are used, then the storage capacity per mass will be reduced [10]. When the short is opened, the stored energy is transferred in part or totally to a load by lowering the current of the coil via negative voltage (positive voltage charges the magnet).

2.2.3 Structure of SMES

The SMES system consists of four main components as they appear in the block diagram from Figure 2.1 [2, 11].

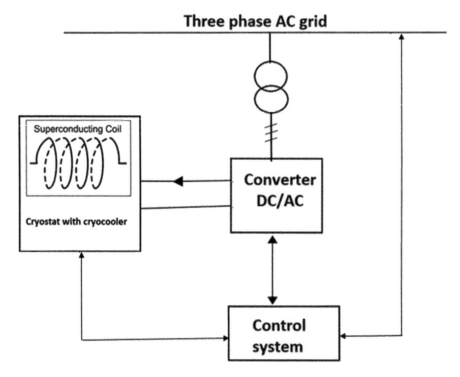

Figure 2.1   Basic architecture of an SMES

- Superconducting magnet with the reinforcement structure
- Cryogenic system
- Power conditioning system (interface between the superconducting magnet and the load or electric grid)
- Control and protection system

The transformer has the role of equalizing the voltages in the network and in the superconducting coil and also of protecting the converter that works at lower voltages than those in the network. The controller monitors and controls the power flow from the network to the coil or from the coil to the network (Figure 2.1).

A rectifier/inverter, is part of the power conditioning system, as required to convert DC of the superconducting coil to AC and vice versa since the very large majority of the grids operate in AC.

2.2.4   *Stress in superconducting coils*

2.2.4.1   **Lorentz forces**

When a conductive element in a magnetic field B is traversed by a current density J, it will develop a Lorentz force density (2.3).

$$\overline{f_L} = \overline{J} \times \overline{B} \tag{2.3}$$

To find the total force, the force volume density must be integrated on the whole volume and obtain (2.4).

$$\overline{F} = \overline{f_v} dv \tag{2.4}$$

The Lorentz forces generate different types of stress in the coil. This is one of the main challenges of the SMES [12–14].

2.2.4.2   **Virial theorem for magnetic energy storage**

The magnetic energy stored in a magnetic system that does not consist of magnetic media can be written as in (2.5).

$$W_m = \int \frac{\overline{B}.\overline{H}}{2} dv = \int \frac{B^2}{2\mu_0} dv = \int w_m dv \tag{2.5}$$

where w_m is the magnetic energy volume density.

It looks like the magnetic flux density values B limits the energy per unit volume given by (2.5). In fact, the real limit of the energy stored in an SMES is mechanical and is given by the Virial theorem [5, 10–12].

The magnetic forces that occur in the coil produce stress. The force and stress distributions depend on the topology.

A. The Virial theorem

(2.6) gives a relation between the mass M of the mechanical structure and the stored energy, W_{mag} [12, 14].

$$\frac{W_{\text{mag}}}{M} = k\frac{\sigma}{\rho} \tag{2.6}$$

Where σ is the maximum working stress, ρ is the mass density, and k is a factor depending on the configuration of the coil. This factor has been calculated for several configurations [5, 12]: $k = 1/3$ for an infinitely thin torus with circular section, $k < 1/3$ for a real toroid with circular section, $k = 1/3$ for an infinitely thin and long solenoid, and $k > 1/3$ for a short solenoid. Practically, $k \approx 1/2$ is reachable.

In order to reduce the high value, Lorentz forces that occur two global ways were proposed: earth- (warm) and self-supported (cold) SMES [11]. In the first approach, the forces are transmitted to the external rock. The magnet is supposed to be installed in an underground cavern or in reinforced surface trenches. In the self-supporting design, the cold structure of the magnet itself supports the Lorentz forces. This option is cheaper than the earth-supported solution for stored energy up to GWh.

There are also other approaches that try to reduce the stress such as force-balanced coils [9, 11, 14–17]. They are helical type hybrid coil that minimize the working stress and thus the mass of the structure. The Virial minimum can be then approached with these topologies, but they remain complex for the winding.

2.2.5   Advantages of SMES systems

SMES is an emerging energy storage technology. For an energy storage device, two quantities are important: the energy and the power. They depend on the application. The energy is given by the product of the mean power and the discharging time. To protect a sensitive electric load from voltage sags, the discharging time must be short (milliseconds to seconds). For load leveling in a power grid, the discharging time should be large (hours to weeks) [11].

The characteristic for SMES is high power, up to 100 MW, and short discharging time under several seconds.

Overall, SMES shows a relatively low energy density.

The energy stored in the superconducting magnet can be released in a very short time. The power per unit mass does not have a theoretical limit and can be extremely high (100 MW/kg). High powers require large currents and an excellent electric isolation for high voltages.

Consequently, the energy with a higher density can be stored in a persistent mode until required.

The main advantages of SMES systems are as follows [11, 17]:

- The current density of SMES coil is about 10–100 times larger than the common coil and has virtually no resistive losses.
  The current density should not exceed a critical value in superconductive materials. It is higher with one up to two orders of magnitude then in conventional conductors. In order to allow the flowing of the same current, the cross section of the copper conductor should be about 100 times larger than that of the superconductor. The resistivity of the superconductors is higher than of the normal metals, at temperatures higher than the critical value [13].
- High power density but rather low high energy density (more a power source than an energy storage device).
- SMES can be easily controlled. With the well-developed power electronic technology, SMES can enhance power system stability and improve the power quality through active and reactive power compensation.
- Very quick response time (up to few seconds).
- Very high number of charge–discharge cycles, being mainly limited by the mechanical fatigue of the support structure.
- No moving parts/low maintenance.
- Fast recharging.
- High energy conversion efficiency (up to 97%).

2.3 Types of SMES

From the point of view of the geometry used in the construction of the superconducting coils, two most classical topologies are used: the solenoid (the multiple solenoids) and the modular toroid. Multiple solenoid type shows very good characteristics on stray field in comparison with solenoid type, but it has very poor energy density [18–20]. Both solutions have advantages and drawbacks. Some of them are manufactured using low-temperature superconductor (LTS) and some using HTS. In many small- and medium-scale SMES systems, the solenoid configuration is preferable because it stores more energy than a toroid SMES using the same amount of superconductors. In addition, the simple structure makes it easier to manufacture. A drawback of a solenoid configuration is the stray field, which is a leakage magnetic field and presents threat to environment and human health. In a toroid configuration, the magnetic field is concentrated within the coil and the stray magnetic field is significantly reduced [18–22]. In Reference 23, a comparative analysis of the stored magnetic energy in an ideal cylindrical and toroidal superconducting magnetic coils is performed. Similar dimensions are used for both coils (diameter and cross section), in analytical and numerical modeling. An HTS Bi-2212 superconductive material was considered for the coils with a diameter of 0.8 mm. In References 24, 25, stress is computed analytically and numerically considering the same geometries but using three different superconducting materials: Nb-Ti, YBCO, and BSSCO. In Reference 26, a comparative study regarding the main storage technologies are presented.

2.3.1 Solenoid configuration

Generally, in literature two cases of this configuration are discussed: thin-walled solenoids and thick-walled solenoids. For thin-walled solenoids, the shape factor (the ratio between the outer and the inner diameter) tends to be 1, whereas for thick-walled solenoids, it is more than 1. Long and thin solenoids are mechanically more stable than thick ones [2, 12, 19, 22, 27].

Because of the high values of the electric current that flows in the coil, a magnetic pressure is exerted on the solenoid walls. The expression of the magnetic pressure can be described with the following expression:

$$p = \frac{B^2}{2\mu_0} \tag{2.7}$$

where B is the value of magnetic flux density, and μ_0 is the magnetic permeability of the vacuum. It represents the magnetic field volume energy density.

The geometry of the solenoid is described by the length l, the radius R, and the number of turns N. The magnetic flux density can be computed using the following expression:

$$B = \mu_0 \frac{NI}{l} \tag{2.8}$$

The self-inductance has the following expression:

Figure 2.2   Solenoidal superconductive coil [27]

$$L = \mu_0 N^2 \Pi R^2 l \qquad (2.9)$$

The stored magnetic energy can be computed by integrating the volume density over the volume of the solenoid.

$$W_m = \frac{LI^2}{2} = \frac{B^2}{2\mu_0} \pi R^2 l \qquad (2.10)$$

A superconductive solenoid coil is presented in Figure 2.2 [27].

2.3.2   Toroid configuration

The expression of the magnetic flux density is

$$B = \frac{\mu_0 NI}{2\Pi} \qquad (2.11)$$

The formula for the self-inductance is obtained from the total magnetic flux.

$$L = \frac{N\Phi_B}{I} = \frac{\mu_0 N^2 h}{2\Pi} \ln \left(\frac{b}{a} \right) \qquad (2.12)$$

The energy density of the magnetic field is represented as follows:

$$W_m = \frac{1}{2}\frac{B^2}{\mu_0} = \frac{\mu_0 N^2 I^2}{8\pi^2 r^2} \tag{2.13}$$

The total energy stored in the magnetic field can be found by integrating over the volume.

$$w_m = \int_a^b \left(\frac{\mu_0 N^2 I^2}{8\pi^2 r^2}\right) 2\pi r h dr = \frac{\mu_0 N^2 h}{4\pi}\ln\left(\frac{a}{b}\right) \tag{2.14}$$

2.3.3 Case study: Computing the magnetic energy storage and the Lorentz force in a solenoid and a toroid

A. Solenoidal geometry

In thin solenoid, Lorentz forces push the coil outwards in a radial direction, creating a hoop stress on the wires. A thin solenoid having the length 123.75 mm, the inner diameter 188.72 mm, and the outer diameter 215 mm was modeled and simulated using the COMSOL Multiphysics finite element method. A Nb-Ti round wire was used for the winding. The current density in the coils is 493 A/mm^2, and the nominal current that flows in the coil is $I = 450$ A [28]. The analytical value of the stored magnetic energy is $W_{anal} = 50,625$ J, and the numerical value is $W_{num} = 53,419$ J.

The magnetostatic field is described by the following equation:

$$\nabla \times (\mu_0^{-1}\nabla \times A) = J_e \tag{2.15}$$

Where J_e is the current density, and A is the magnetic vector potential.

The magnetic pressure (magnetic field volume energy density) is represented in Figure 2.3.

When a conductive element is traversed by a current density J in a magnetic field B, it will develop a Lorentz force density having the following expression:

$$\overline{f_L} = \overline{J} \times \overline{B} \tag{2.16}$$

To find the total force, the force density should be integrated on the volume.

$$\overline{F} = \overline{f_v}dV \tag{2.17}$$

The Lorentz force distribution is represented in Figure 2.4. Also, the vector orientation is presented. The force has a radial distribution that tends to extend the circumference of the coil.

Integrating on the contour of the coil, we obtain the values of the Lorentz force axial and radial components: $F_{Lz} = 0,4$ and. $F_{Lr} = 8.10^5$ N

The high value of the radial force generates a von Misses stress. In order to determine the stress distribution, a coupled magneto-structural problem has to be solved. The output of the magnetostatic problem, the Lorentz force is applied as an input to the structural problem, in order to find the von Misses stress.

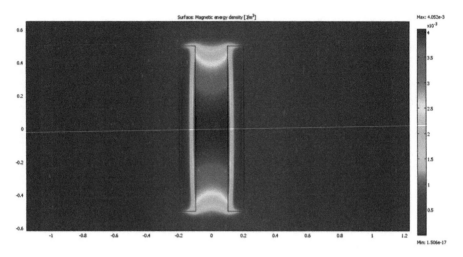

Figure 2.3   Distribution of the magnetic pressure in an ideal solenoidal coil

In Figure 2.5, the distribution of the von Misses stress for a solenoidal coil made of Nb-Ti and the mid-plane deformation can be noticed.

The maximum value of the stress is less than the yield value for the Nb-Ti material. Hence, the stability of the material is confirmed.

B. Modular toroidal geometry

In Figure 2.6, the modular coils are inside a cube and the 3D domain is discretized with tetrahedral finite elements. In Figure 2.7, a detail is presented. It was modeled using the same COMSOL Multiphysics finite element software package. The structural parameters of the toroidal magnet are the number of element coil

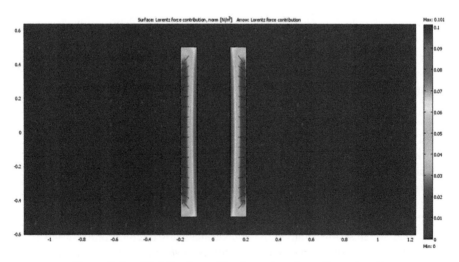

Figure 2.4   Distribution of the Lorentz volume force density

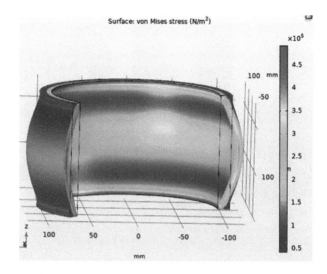

Figure 2.5   von Misses distribution for an ideal solenoidal coil

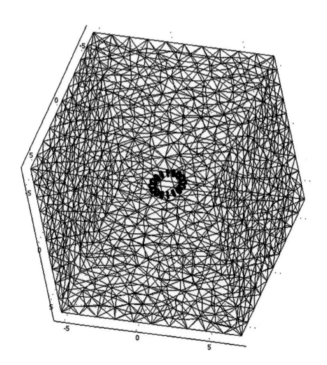

*Figure 2.6   Wireframe representation of the finite element discretization of the
cube domain with the modular toroidal coil*

Figure 2.7   Finite element discretization of the modular toroidal coil

(*N*), the internal radius of element coil (*R_i*), the external radius of element coil (*R_o*), the height of element coil (*H*), and the distance from element coil center to toroidal magnet center (*R*) [19, 21, 29]. There are *N* = 16 element coils. The structural parameters of toroidal magnet have a great effect on storage energy. It is known that stored energy increases with the increase in number of element coil (*N*). Because when number of element coil (*N*) is added, the greater coupling between element coils will be obtained, which tremendously increases the storage energy. Consequently, the larger number element coil is beneficial to the large capability toroidal SMES magnet.

In Figure 2.8, the magnetic flux distribution of the modular toroidal coil is described.

2.3.4.   Comparison between solenoidal and toroidal geometries

Different comparisons between them are presented in References 18, 19, 30, 31.

In Reference 32, a design of a 2G HTS SMES coil with solenoidal and toroidal configurations having a power rating of 1 MW for a duration of 5 s is presented. The comparison is performed considering conductor/cable choice, dimension of the magnet bore and AC losses using 2G HTS YBCO tape in the form of a Roebel cable [31]. First ideal toroidal and solenoidal configurations are analyzed, and the technique used to choose the bore of the coils is described. The same amount of stored energy, 5.6 MJ, was considered and a 1.25 H value for the coil inductance was used. The stray field was compared using 2D models. Lines of constant stray field at 500, 100, and 50 mT for the solenoid were presented, and the concentration of the magnetic field in the toroidal coil was emphasized. The total length of HTS tape for the toroid and solenoid was 59.8 and 68.1 km, respectively. The losses were compared for the same field level and slightly higher losses per cycle per unit length of conductor appeared in case of the toroid, albeit the higher transport current of 300 A for the toroid comparatively with a 188 A for solenoid. This suggests that the losses

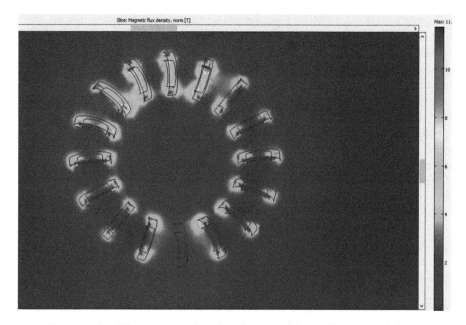

Figure 2.8   The magnetic flux distribution of the modular toroidal coil

are dominated by the applied field [33]. Unlike a hybrid solenoid magnet discussed above, the same toroidal geometry of a 2.5 MJ SMES by either rare earth barium copper oxide (REBCO) or bismuth strontium calcium copper oxide (BSCCO) conductors was also reported to show that YBCO may provide advantages of smaller magnet size, less total conductor length and smaller AC losses, but the high magnetic flux density in YBCO toroid may lead to very high mechanical stresses [33].

In Reference 34, a 2.5 MJ class SMES with HTS magnets of single solenoid, multiple solenoid, and modular toroid type were optimized using a recently developed multi-modal optimization technique named multi-grouped particle swarm optimization. The objective of the optimization was to minimize the total length of HTS superconductor wires.

Various configurations of HTS magnet and design variables were considered: single solenoid, two solenoids, four solenoids, and a toroid. Stored magnetic energy, stray magnetic field, and critical magnetic field were computed using the finite element discretization with tetrahedral elements. The conclusion was that in order to store 2.5 MJ, the toroid type requires the minimal total length of HTS wire by increasing operating current and has the good characteristics about the minimization of stray field. In Reference 35, both solenoid and toroid (composed of eight identical coils) configurations are designed and compared.

The capacities of SMES systems are designed on different scales including kJ-, MJ-, and GJ-class for different applications. Finally, the costs are calculated based on the commercial price of superconductors. The result indicates that the solenoid configuration is suitable for kJ-scale SMES, while the toroid configuration is suitable for GJ-scale SMES. For a small-capacity superconducting magnet, the solenoid-type configuration will be the prime consideration. The simulation result shows that a higher value of the radius helps to reduce the high magnetic field and increase the storage capacity. However, more superconductors are required with the increase in the coil radius. Applying a larger current into conductor is an efficient way to increase the stored energy. However, the maximum possible current is limited by the magnetic field produced by the current. Therefore, it is necessary to find a balance between increasing energy capacity and ensuring stability. Two values of the stored energy were considered for each configuration 1.2 KJ and 1.6 MJ for the solenoid and 1.3 MJ and 1 GJ for the toroid. The obtained total lengths of superconductors tape were 0.1 and 12 km for the solenoid and 5 and 533 km for the toroid, respectively. For a medium or large SMES system, a toroid configuration is a better option.

The modeling result indicates that a greater energy capacity could be achieved by a bigger radius of the toroid magnet. The increase in the radius could also reduce the magnetic flux density at the same time. However, more supporting structure materials and cryogenic costs are required.

2.4   Dynamics models of SMES

In an electrical network, storage systems play an important role because they help maintain the balance between production and energy consumption [36, 37]. An SMES has the ability to go from full charge to full discharge very quickly, which would make it extremely useful for integration with renewable technologies to mitigate the adverse impacts. A suitable core needs to be designed to allow SMES to have high stored energy. The dynamics of these storage devices is represented by the operating modes. More precisely, an SMES operates in three distinct modes: charge, storage, and discharge. The total efficiency of an SMES system can be very high since it does not require energy conversion from electrical to mechanical or chemical energy. Depending on the control loop of its power conversion unit and switching characteristics, the SMES system can respond very rapidly (MWs/milliseconds). The ability of injecting/absorbing real or reactive power can increase the effectiveness of the control and enhance system reliability and availability. Consequently, SMES has inherently high storage efficiency of about 90% or greater round-trip efficiency.

References 37–39 present elements related to the dynamics of the energy storage process, for several power storage devices as well as SMES. This notion of dynamics related to energy storage refers to the way of managing the current circulation through the storage devices to and from the electrical network in conditions of stability and safety.

In Reference 40, a general dynamic model for SMES using a pulse width modulation (PWM)-current source converter (CSC) power converter to connect the grid is presented. The difference from the methods presented above is that the dynamic model of SMES using PWM-CSC is described by a set of equations in

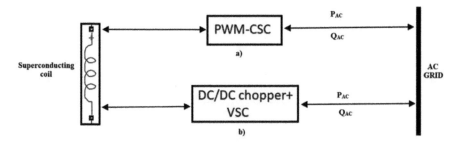

Figure 2.9   Configurations for SMES integration adapted from Reference 40

d–q coordinates. These are obtained by applying Kirchhoff's theorems in the part comprising the transformer and the network and Tellegen's theorems for calculating the power transfer from the AC part to the DC part. The presentation of the main advantages for each control method is mentioned below.

- Thyristor-based SMES: The thyristor-based SMES can control mainly the active power and has a little ability to control the reactive power. Also, the controls of active and reactive powers are not independent. The thyristor-based SMES is easier to control, having only one AC/DC module.
- Voltage source converter (VSC)-based SMES: The VSC-based SMES allows an independent control of the active and reactive power flowing between the superconducting coil and the power system network. The VSC-based SMES includes not only an AC/DC circuit but also a DC/DC chopper. Thus, the control is complicated compared to both the thyristor- and CSC-based SMES.
- CSC-based SMES: The CSC-based SMES allows an independent control of the real and reactive power flowing between the superconducting coil and the grid. Having only one AC/DC module, the CSC-based SMES is easier to control.

In Figure 2.9, the last two cases are described.

The most used topology configuration of power conversion systems (PCSs) for SMES are VSC- and CSC-based PCS.

In Reference 41, two types of PCSs are presented: a PCS based on thyristor and a VSC using insulated gate bipolar transistor (IGBT). In the first case, the control of the energy flow is done in a relatively simple way by changing the control angle of the thyristors. If α is considered the control angle, the following situations are encountered: in the charging process, for $\alpha = 90$, the current through the coil remains constant in the storage process, and for $\alpha > 90$, the energy in the coil is released in the network. The PCS using thyristors are made using six-pulse converter or twelve-pulse converters [41]. The VSC requires a six-pulse PWM rectifier (or) inverter using IGBT, a two-quadrant DC–DC chopper using IGBT. The PWM converter and the DC–DC chopper are linked by a DC link capacitor [42]. The connection to the

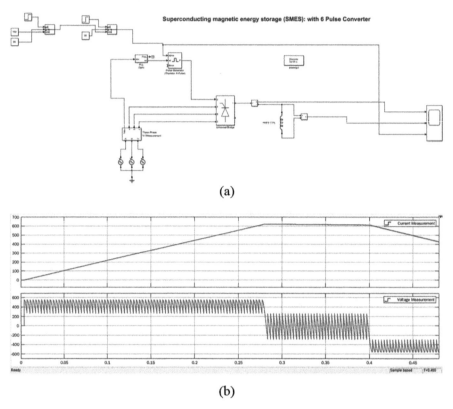

(a)

(b)

Figure 2.10   Control of superconducting coil six-pulse: (a) Simulink® diagram and (b) results of simulation

utility grid is made by means of a step-up Δ–Y coupling transformer, and second-order low pass sine wave filters are included in order to reduce the perturbation on the distribution system from high-frequency switching harmonics generated by the PWM. Since two ways for linking the filter can be employed, i.e., placing it before and after the coupling transformer. Here, it is preferred the first option because it reduces notably the harmonic contents into the transformer windings, thus reducing losses and avoiding its overrating [42, 43].

For the numerical study of the dynamics of an SMES, a model is made with the help of MATLAB-Simulink® software. A three-phase alternating voltage source is used to make the diagram, which supplies a superconducting coil with a 0.2 H inductor via a thyristor bridge [44]. The thyristor bridge is controlled by the angle α. For different values of this angle, the coil enters different processes such as charging, discharging, or energy storage. Figure 2.10 (a) shows the diagram made in Simulink® in which the coil is subjected to three processes. Initially, for loading, the thyristor

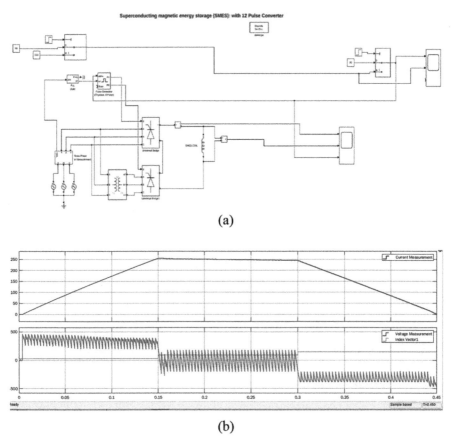

(a)

(b)

*Figure 2.11   Control of superconducting coil 12-pulse: (a) Simulink® diagram
and (b) results of simulation*

bridge receives a control angle of 30 degrees. For the loading process, the angle is
changed to 90 degrees and for unloading its value α is 150 degrees. Figure 2.10 (b)
shows the values of voltage and current on the coil following the simulations. As
for the time intervals for which the coil can be in one of the processes, it depends on
the value entered in the step signal block that makes the transition from one control
angle to another.

For a better voltage recovery because the SMES operates in direct current, a
second thyristor bridge connected in a triangle with the first can be used. Such a
model which is shown in Figure 2.11 (a) and Figure 2.11 (b) shows the values for
the voltage and current flowing through the SMES perform the transition from one
control angle to another [41].

Also, in the application of high power, the CSC has an additional advantage,
that is, being easily paralleled of multiple bridges [45].

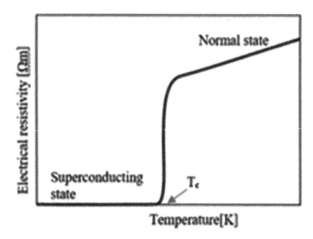

Figure 2.12   Transition from normal conduction state to superconducting state

2.5   Materials, challenges, and cooling system of SMES

2.5.1   *Basics of superconductivity*

The resistivity of some metals and alloys, cooled below a specific low temperature, drops rapidly to zero. This phenomenon, known as superconductivity, was first observed by Dutch physicist Heike Kamerlingh Onnes. In 1908, Kamerlingh Onnes liquefied helium at a temperature of 4.2 K and in 1911, he discovered the superconductivity of mercury at 4.2 K [46–48].

In Figure 2.12, the transition from normal conduction state to superconducting state is represented and described by the sharp decrease in the electric resistivity versus the temperature. In Figure 2.13, a comparison between the resistivity of a normal metal and a superconductor is depicted.

All superconductors must operate within a regime bounded by three inter-related critical quantities: current density, operating temperature, and magnetic field. The highest temperature at which material has no electrical resistivity, when the superconductive state occurs, is called critical temperature (T_c). The upper limit of its current carrying capability is called the critical current density (J_c) and critical magnetic field (H_c). Beyond those values, the superconductive state is cancelled [48].

The superconducting state of a material is described by its critical surface describing the critical current density J_C dependence versus the temperature T and the magnetic flux density B.

The generic critical surface is presented in Figure 2.14 and a comparative critical surface for Nb-Ti and Nb_3Sn superconductors is presented in Figure 2.15

the operating critical temperature and field are below the critical values, the transition superconducting to normal state is possible if the current inside the superconductor is higher than I_C. Between these two states, the transition is called dissipative

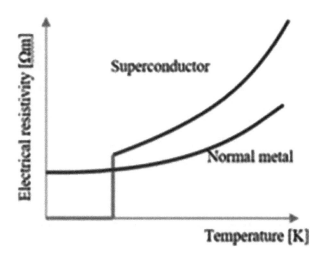

Figure 2.13   A comparison between the resistivity of a normal metal and of a superconductor

state. This state describes the penetration of the current in the superconductor and is important for numerical modeling. (Figure 2.15)

When the current in a superconductor exceeds the value of the critical current, an electric field occurs. The relation *E–J* can be evaluated using experimental characterizations. Dependence of the electric field versus current can be described by a "power law," having the following expression [48–50]:

$$E = E_c \left(\frac{|I|}{I_c} \right)^n \cdot \left(\frac{I}{|I|} \right) \tag{2.18}$$

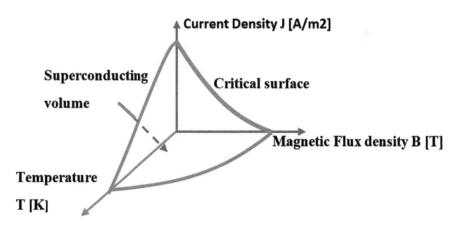

Figure 2.14   The generic critical surface of the superconducting state

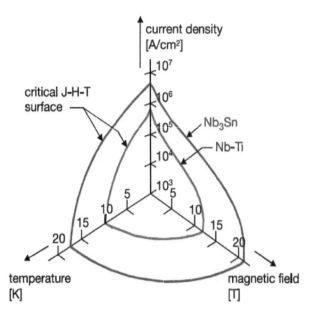

Figure 2.15   A comparative critical surface for Nb-Ti and Nb$_3$Sn superconductor [47]

where E_C is the critical field and is chosen to be 1 μV/cm (international standard for HTS). This value is obtained from experimental results [51]. The power exponent "n" is the resistive transition index. The value of n depends on the material, temperature, and magnetic field. For Nb-Ti or Nb$_3$Sn at 4.2 K, the n values are about 100 while for REBCO conductors, at the same temperature n values are about 30 [50].

2.5.2   Superconductor used in electrical engineering applications (SMES systems)

There are two types of superconductors, type I and II, depending on their critical magnetic field [49]. Mostly type II superconductors are used in power applications. The other superconductors discovered were pure metals like lead and tin, which had practically zero current carrying capability in the presence of high magnetic fields. However, in the late 1950s and early 1960s, a new class of high-field alloys like niobium-titanium (Nb-Ti) and compounds such as niobium-tin (Nb$_3$Sn) were discovered. Table 2.1 presents a comparison of basic material characteristics and critical current density information for superconductors. These superconductors, operating at about 4.2 K and the liquid helium boiling point at room pressure, are called LTS.

Nb-Ti (niobium titanium), with a critical temperature of 9.5 K, is mostly used in large-scale applications. It is relatively easy to manufacture even in long lengths and it has good mechanical properties. It covers 80% of the superconductors market (6 billion dollars in 2017, about 3,000 tones/year). At 4.2 K provides a magnetic

Table 2.1   Basic material and critical current density relevant parameters for practical superconductors and their wire fabrication technology and typical forms at present [48]

Material	$T_{c,max}$ (K)	Hc2, 4.2 K (T)	J_c 4.2 K (A/cm²)	Coherence length, ε_{ab} (nm)	Anisotropy, γ_H	Wire technology	Typical wire forms
Nb-Ti	9.5	11.5	4×10^5 (5T)	4	Negligible	\	Round wire
Nb₃Sn	18	25	$\sim 10^6$	3	Negligible	Bronze process internal Sn process powder-in-tube	Round wire
MgB₂	39	18	$\sim 10^6$	6.5	2–2.7	Powder-in-tube internal Mg diffusion	Wire or tape
REBCO	92	>100	$\sim 10^7$	1.5	5–7	Coated conductor	Flat tape
Bi-2223	108	>100	$\sim 10^6$	1.5	50–90	Powder-in-tube	Flat tape
Bi-2212	90	>100	$\sim 10^6$	1.5	50–90	Powder-in-tube	Wire or tape
1 111 IBS	55	>100	$\sim 10^6$	1.8–2.3	4–5	Powder-in-tube	Wire or tape
1 22 IBS	38	>80	$\sim 10^6$	1.5–2.4	1.5–2	Powder-in-tube	Wire or tape
1 1 IBS	16	>40	$\sim 10^5$	1.2	1.1–1.9	Powder-in-tube	Wire or tape

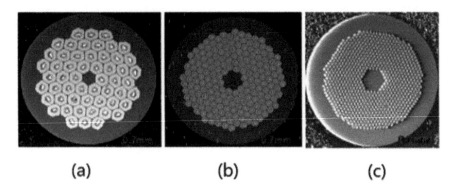

(a)   **(b)**   **(c)**

Figure 2.16   Cross section of Hyper Tech 0.7 mm diameter "tubular" approach
Nb₃Sn wires with the following number of filaments: (a) 61, (b) 217
and (c) 919 (Courtesy: Hyper tech Research Inc.)

flux density of 10 T and at 1.8 K of value of 12 T [50, 51]. Basic material and critical current density relevant parameters for practical superconductors and their wire
fabrication technology and typical forms are described in Table 2.1.

Also, LTS Nb₃Sn is used in power applications. The production process is more
difficult than Nb-Ti, but it has higher critical temperature (18 K) and critical magnetic flux density. At 4.2 K, Nb₃Sn have a critical field of 25 T. Both of them have
current densities over 10^5 A/cm², which are about two orders of magnitude higher
than that of copper conductors. Consequently using these materials high values of
magnetic fields are obtained by comparison with conventional magnets [47].

In Figure 2.16 (a–c), certain cross sections of Nb₃Sn wires, with different number of filaments, are presented.

In Figure 2.17, cross sections of 0.7 mm (left) and 0.45 mm (right) diameter
"tubular" approach Nb₃Sn 919 filament showing filament size of 16 and 10 are
depicted.

In Figure 2.18, a conduction cooled Nb₃Sn persistent coil with persistent switch
and integrated persistent joints – soon to be tested at Ohio State University – is
described.

A. High-temperature superconductors

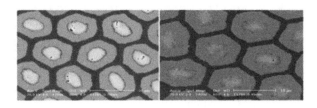

Figure 2.17   Cross section of 0.7 mm (left) and 0.45 mm (right) diameter
"tubular" approach Nb₃Sn 919 filament showing filament size of 16
and 10 (Courtesy: Hypertech Research Inc.)

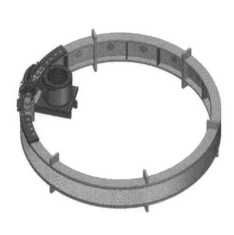

*Figure 2.18   Conduction cooled Nb$_3$Sn persistent coil with persistent switch
and integrated persistent joints – "Soon to be tested at Ohio State
University" (Courtesy: Hyper tech Research Inc.)*

They appeared in 1986 when J. Bednorz and K. Muller discovered LaBaCuO
superconductors having a critical temperature T_c of 35 K. In 1987, the T_c exceeded
for the first time the liquid nitrogen (LN) temperature (77 K) by the discovery of
YBCO or REBCO, RE = rare earth superconductors with T_c up to 93 K. They can
be cooled with LN for applications in low magnetic fields. They were used with
cryocoolers for high magnetic fields in the range of 30–40 K [49, 50].

In 1987, the bismuth-based cuprate superconductors (BSCCO), including Bi-
2212 and Bi-2223 with T_c up to 110 K were discovered.

Bi-2223 and REBCO can carry high currents being useful not only for high-field
magnets operated in low or moderate temperature range but also for electrical applications at higher temperature that use LN as coolant. Even if Bi-2212 can be used
at low temperatures (<20 K), it has certain advantages for high-field applications.

The BSCCO-2212, BSCCO-2223, and MgB$_2$ (considered a medium or intermediate temperature superconductor, MTS) are considered first-generation (1G) high
temperature superconductors. These superconductors represent the perspective of
increased efficiency and low operating cost for the power applications.

However, only BSCCO-2223 and YBCO-123 wires capable of operation in the
temperature range of 20–70 K have achieved widespread application for manufacturing practical electric power equipment.

These applications employing LTS materials are very sensitive to even a small
thermal energy injection that could raise local temperature sufficiently to drive the
magnet into quench.

Very low heat capacity of materials at low temperatures (around 4 K) causes this phenomenon. The HTS materials in conductors operating at about 30 K have heat capacity hundred times higher than 4 K. Thus, any local energy injection causes a much smaller temperature rise. HTS materials also transition slowly to their normal state because of a low resistive transition index n.

The previous aspects enable the HTS magnet to operate successfully in the presence of significant local heating. HTS materials have proved to be successful in many applications where LTS materials have been unsuccessful especially for industrial applications [49].

HTS conductors operating at LN temperature are employed for building power cables, fault current limiters (FCL), and transformers for applications in AC electric grid.

1. BSCCO-2212 superconductors

BSCCO was the first HTS material used for manufacturing practical superconducting wires. The critical temperature (T_c) of BSCCO-2212 is about 90 K and has high magnetic field features in the temperature range between 4.2 and 20 K. This material has a very low current carrying capacity in a magnetic field at 77 K. It was the first HTS material used for making superconducting wires and is very versatile. BSCCO-2212 can be manufactured as round wire and flat tape with many dimensions. The critical current in the round wires shows no anisotropy with respect to an applied field.

This material has been used to manufacture long conductors with uniform properties. BSCCO-2212 offers a potential to use in engineering applications for operating temperature lower than 20 K but at temperatures higher than 20 K the performance is limited.

The critical current density (J_c) versus the magnetic flux density B decreases drastically at higher temperatures.

BSCCO-2223 HTS has a critical temperature of about 110 K, which is about 20 K higher than BSCCO-2212. It was manufactured in tape lengths of several hundred meters. This 1G HTS wire has a tape shape, typically 0.2×4 mm, consisting of 55 or more tape-shaped filaments, each 10 μm thick and up to 200 μm wide, embedded in a silver alloy matrix.

In Reference 50, the applications of BSCCO-2223 and BSCCO-2212 to a high-temperature SMES system are described. The system was designed and realized by ACCEL Instruments GmbH together with AEG SVS GmbH, EUS GmbH and E.ON Bayern AG companies. The magnet is designed for an energy of 150 kJ.

The superconducting coil is composed of an HTS tape with BSCCO-2223 filaments. The current leads are composed of BSCCO-2212 element. The system will be included in a 20 kVA uninterruptible power supply system coupled to the electrical grid for enhancing the power quality of a selected grid user.

2. 2G High temperature superconductors

In 2001, the superconductivity at 39 K in MgB_2 was discovered by the Akimitsu group in Aoyama-Gakuin University, Japan. MgB_2 can be used in applications at temperatures about 20 K that can be obtained by liquid hydrogen or cryocoolers. It seems that they replace Nb-Ti and Nb_3Sn used in liquid helium [49–51].

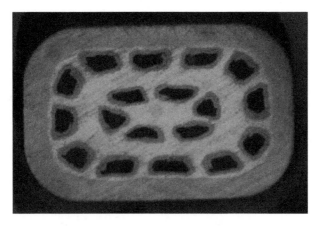

Figure 2.19   MgB$_2$ wire for MRI application (Courtesy: Hyper tech Research Inc.)

The light weight of MgB$_2$ makes it attractive for large-scale practical applications. It is produced at kilometer-level practical MgB$_2$ wires by Hyper Tech in the USA, Columbus in Italy, Hitachi in Japan, Sam Dong in South Korea and Western Superconducting Technologies in China. In the next section, certain images with MgB$_2$ wire and coils offered by Hyper Tech Research Inc. from USA are presented.

In Figure 2.19, a cross section of a MgB$_2$ wire for MRI application is presented.

In Figure 2.20, some cross sections through AC tolerant MgB2 wire for cable are presented.

In Figure 2.21, an SMES coil wound with 2.1 km of MgB2 wire is presented.

3.YBCO-123-coated superconductors

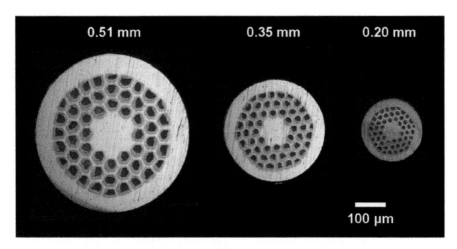

Figure 2.20   AC tolerant MgB$_2$ wire for cable (Courtesy: Hyper tech Research Inc.)

Figure 2.21   SMES coil wound with 2.1 km of MgB$_2$ wire (Courtesy: Hyper tech Research Inc.)

The 2G of HTS materials based on thin films was yttrium barium copper oxide YBa2Cu3O7 or YBCO-123. The YBCO-123 in the form of coated conductors has also advanced significantly to provide current density capability suitable for application in practical devices. Roebel cables, made from the coated conductors, could carry currents in kA range with minimal losses. Because of material characteristics of BSCCO-2223-based HTS material, 1G wires were limited to temperatures lower than 40 K while operating in magnetic fields above 2 T. At 77 K, the 2G wire retains its J_c at much higher magnetic fields than the 1G wire.

4. Iron-based superconductors

In 2008, the iron-based superconductors were discovered by the Hosono group at the Tokyo Institute of Technology. Like the cuprate superconductors, iron-based superconductors have high critical fields and electromagnetic anisotropy. The critical temperature of iron-based superconductors is between 38 and 56 K. These properties make iron-based superconductors attractive for applications that use liquid helium and also around 20 K [48, 50].

In order to obtain a wide application of this technology, a combination of low-cost HTS conductors and refrigeration systems could be a promising one [50, 51].

2.5.3.1   Challenges of superconducting materials

In order to be used in power applications, superconducting materials must have high T_c and transport high electrical currents in a high magnetic field. They should have stable mechanical properties and low-cost production [49, 50].

High-field magnet applications require the usage of HTS superconductors forcing the extending of their market. But LTS materials have better prices, a mature technology, and high stability. All these advantages make them the main option for applications, where the necessary temperature and field strength cover their performance range.

In order to develop large-scale applications of HTS BSCCO, MgB_2 and REBCO performance/cost ratio should be increased. HTSs are also promising materials to realize efficient energy storage devices [51, 52]. Their main advantage is the high superconducting transition temperature T_c compared to LTSs. This decreases the cooling costs, a very important aspect for large-scale applications.

The dream of researchers is the room temperature superconductivity. Recently, some achievements in this direction were reported [53, 54]. A carbonaceous sulfur hydride with a critical temperature up to 287.7 K (15 C) under an extremely high pressure of 267 GPa was obtained [55,56].

But the operation at such high pressures is much more difficult than that at low temperatures. This result will stimulate the search for practical room temperature superconducting materials in the future [50, 51].

A superconducting wire carrying a current that never vanishes represents a perfect SMES device operating at room temperature. Unlike conventional batteries, which degrade over time, in a room temperature SMES device energy can be stored for a long time without any appreciable losses.

2.5.3.2 Cooling systems of SMESs

The main component of the SMES is the superconducting coil. In order to store energy without loss and to maintain the superconducting state of the coil, this superconducting coil must be cooled with a cryogenic cooling system below the critical

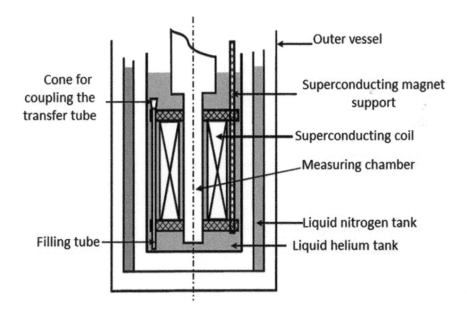

Figure 2.22   Diagram of a cooling system of a superconducting magnet [34]

temperature [57]. HTS superconducting wires such as NbTi and Nb3Sn [58, 59] or REBCO and BSCCO are made of tapes. [57]. Figure 2.22 shows the general structure of a cooling system for a superconducting magnet. The superconducting cylindrical magnet is suspended, by means of supports, from the upper flange of the cryostat that also contains a variable temperature chamber in which the experiment is performed (but there are many applications that do not use such a chamber but one at room temperature).

Since most LTS magnets made with niobium-titanium (Nb-Ti) and niobium-tin (Nb$_2$Sn) superconductors require operation close to 4 K, liquid He has been the only fluid of choice. Almost all high-field LTS magnets employ liquid He as a coolant. However, HTS conductors (BSCCO-2223 and YBCO-123) can operate at LN temperature (77 K) for low-field applications and at liquid He temperature (27 K) for high-field applications. LN is preferred as a coolant because it is inexpensive, inert, and readily available. Liquid He is also attractive as a coolant for HTS magnets but is expensive and not readily available in large quantities. The main disadvantage of all cryogens is that they must be replenished or be used in a closed cycle where a cryocooler recondenses boiled-off vapor. As a point of caution, all established safety and control rules must be followed while working with cryogens because of the hazards associated with them [60].

A. Cooling methods
1. Direct cryogenic cooling

Superconducting magnets that use NbTi and Nb3Sn superconductors are usually cooled by immersing them in liquid He. As the NbTi conductor is introduced into the liquid He, the heat that appears in the conductor is transferred to the coolant. This thermal load is absorbed by the transformation of liquid He into gaseous He using the latent heat of vaporization. In these magnets, the superconductors remain in the superconducting state as long as the heat flux (heat per unit area) available at the surface of the conductor is less than what liquid He can remove; this limit is called the critical heat flow. After the critical heat flow is exceeded, the superconductor temperature also begins to rise, eventually the superconductor enters its normal state when its temperature exceeds its critical temperature T_c.

2. Indirect or conduction cooling

The vast majority of high field magnets have been built using simple and convenient conduction cooling with cryocooler refrigerators. These magnets, which sometimes exceed 12 T, are now available to buyers in the commercial environment. Superconducting coils are impregnated with epoxy to create mechanically durable and easy-to-handle monolithic structures. Such coils are cooled from their outer surface by conduction. Two cooling methods are shown in Figure 2.23, where good thermal contact is implemented between the HTS coils and the refrigeration system: the first method applies when the tubes carry coolant (1) and the second which is directly by a refrigerator cooler (2). The coolant in the cooling tubes can be a cold gas or a liquid cryogen. The coolant can operate in a closed cycle, repelling its thermal charge outside the magnetic system in a refrigerator or other equivalent cooler. The coolant could also be a cryogen taken from a storage container. In addition, a

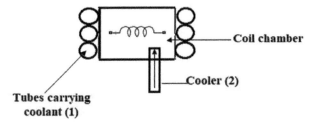

Figure 2.23   *Conduction cooling of a superconducting coil: (1) coolant in tubes and (2) coil in contact with a refrigerator cooler adapted from Reference 47*

cryocooler refrigerator could be thermally interfaced with an HTS coil that is cooled [47, 61].

3. Refrigeration systems

Another method for cooling superconducting magnets is the refrigeration process. The process involves removing a quantity of heat from an enclosed space or a substance to maintain a low temperature. In Table 2.2, the most used refrigerators, Gifford-McMahon (G-M), Stirling, and pulse tube, are presented.

4. Temperatures from various cooling liquids

The study by Green [62, 63] presents the properties of three liquids used in cooling superconducting magnets, namely helium hydrogen and nitrogen. A comparison of temperature range and associated enthalpy range is also presented. Hydrogen has 590 J/ g at an operating temperature range (20–70) K, helium 383.72 J/g at a temperature range (4–77) K, and nitrogen 233.5 J/g at an operating temperature range (77–300) K. It should be noted that hydrogen cooling involves a lower cost in cooling superconducting magnets. In Reference 64, liquid helium is presented as a cooling agent at a temperature of 4.2 K of a superconducting magnet. An analysis of the heat flow distribution and of the heat losses is presented.

5. The cost of cooling systems

Estimation of the cost of a superconducting coil system and the cooling system needed to keep it cold is often difficult. One method for making a cost estimation for two components (superconducting coil and cooling system) is based on knowing what these components have cost in the past. One of the difficulties with this kind of estimate is the choice of the appropriate scaling parameter [65]. The study by Nuno *et al.* [66] presents an analysis of the cooling cost of a superconducting system depending on the desired temperature level and the cooling process used. Another way of analyzing the cost of making an SMES is the one in which a cost analysis is made according to the stored magnetic energy. Reference 64 presents a cost analysis according to the energy that is stored in a superconducting device. This price includes both the cooling cost of the device and the realization cost of the superconducting coil. In case of using a cylindrical coil for a total stored energy of 1.2 kJ, the price is 1.4 $/J, and for a toroidal coil the price for a total stored energy of

Table 2.2 *Refrigerator types and its characteristics [49, 62]*

Refrigerator type	Gifford-McMahon (G-M)	Stirling coolers	Pulse-tube coolers
Characteristics	G-M cryocoolers are widely used for cooling magnets from 2 to 80 K. These coolers are made as one-, two-, and three-stage devices and are made in various sizes. G-M coolers are quite portable and versatile for a large variety of applications and have made possible many HTS-based devices	The Stirling chiller works by repeatedly heating and cooling a sealed amount of working gas, usually helium for cryogenic temperatures. A piston varies the volume of working gas, and the transport of the gas is handled by a device, from the cooler between the hotter components and the colder components	The advantage of a pulse-tube cooler is that the "displacer" is a column of gas, not solid material; it is a gas plug. This eliminates a crucial moving part at low temperatures and greatly enhances reliability and reduces vibration. Current pulse-tube development efforts are concentrated on the basic understanding of the cooling principle, the extension of the temperature range, and the exploration of various cooling arrangements

1.3 MJ is 0.33 $/J, about 4 times smaller than the cylindrical coil. One cause of the lower price in the case of the toroid is a lower flow of leaks.

2.6   SMES applications to power and energy systems

The introduction of electricity to world started about one century ago and since then electrical devices and systems have been widely used [67]. Due to the deterioration and aging of the electrical devices in existing systems, generation and transportation of a clean energy have been a challenge. Since the superconductors have been discovered a large amount of dissipated energy is saved, in cooperation with conventional materials, the superconductors have the potential to bring a fundamental change to power technologies. The first commercial application of an SMES system was developed in 1980–1981 in Los Alamos Laboratory, USA. It consisted of a 30MJ Nb-Ti unit, with a maximum power output of 10 MW, used for low frequency stabilization (0.35 Hz) [67, 68]. The application of superconductors can be divided in two major categories. The first is to use SMES in classical topologies for energy production and distribution such as superconducting cables and superconducting transformers, and for maintaining the grid parameters such as frequency, power, and load fluctuations. Another application of SMES is to store energy from renewable sources like sun and wind.

A. Applications of SMES in classical grids

Modern power transmission systems must mainly provide stability in operation. Therefore, devices are needed to "enter" the system quickly, in case of loss of synchronization of generators with the network, or when major power fluctuations occur. The basic applications of SMES in conventional grids are described below [69].

1. Power system stability

The power stability of the systems is characterized by frequency oscillations in the range (0.5–1) Hz. These frequency variations can lead to unintentional tripping of electric machines, leading to major disturbances. Interventions on both active and reactive power can be much more effective, and smaller in size, than other technologies [70, 71].

2. Improving voltage stability

When there is a loss of active power in the electrical network, a dynamic instability occurs. SMES is used to stabilize the energy system and can operate for a long time, until other energy sources intervene in the system. The schematic diagram of the power control system with the SMES unit for improving voltage stability is similar to Figure 2.24. The generator gives the grid active power P_g and reactive power Q_g. The grid needs power P_c and Q_C. When the power demand of the grid is higher than the power of the generator, the superconducting coil supplies more power the grid.

3. Spinning reserve

Because superconducting systems can store significant amounts of energy, they are used as an energy reserve. A considerable advantage over other storage methods

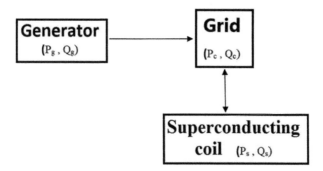

*Figure 2.24   Application of SMES for voltage stability adapted
from Reference [68]*

is the short time it can supply energy to the system. Providing "spinning reserve" with SMES is much more efficient since it is a virtually lossless form of storage, whereas providing spinning reserve with generation has significant losses and high operating costs [72, 73].

4. Improving FACTS performances

SMES systems can be configured to provide energy storage for flexible AC transmission system (FACTS) devices. Kumar and Savardekaro [74] discussed the power quality benefits for transmission systems by integrating FACTS controller with SMES. An inverter based in damping dynamic oscillations is integrated with an SMES into an in-power system. Their studies showed that, depending on the location of the SMES, simultaneous control of active and reactive power can improve system stability and power quality of a transmission grid.

5. Superconducting cables

Superconducting cables have significant advantages of compactness and larger power transmission capacity that is expected to be installed in a power grid to replace the existing power cables. The 1G wire consists of a composite fine filaments of HTS BSCCO-2223 embedded in a silver matrix. The HTS power cable consists of YBCO wire with coated conductor [75].

6. Superconductive transformer

The use of superconducting transformers reduces the risk of fire by removing the electrical insulating oil from its tank. This reduces the risk of pollution and the risk of explosions caused by overpressure of the cooling oil. A superconducting transformer has the capability of short-circuit current limitation under fault conditions. It can work continuously in overloaded conditions without insulation damages and any lifetime loss because of the ultra-cold operating environment [76].

7. Electromagnetic launcher

This type of application is used to convert electrical energy into mechanical energy to achieve a linear motion of a projectile at a very high speed. An electromagnetic launcher requires current higher than 1 MA, and it can launch projectiles at speed 2,000 m/s, surpassing conventional methods [74]. The components of an electromagnetic launcher are two rails connected to a pulse current generator,

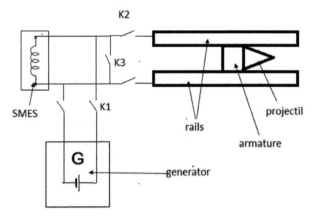

Figure 2.25   *Operation of an electromagnetic launcher adapted from Reference 11*

a conductive armature, and a projectile. SMES can be used as an energy storage device for an electromagnetic launcher due to its high power density. The current flows through the rails and armature and creates a magnetic field, which causes an electromagnetic force on the armature due to Lorentz force [11, 74–76].

(1) describes the expression of the Lorentz force acting on the moving armature.

$$\overline{F}_{arm} = \int \overline{J} \times \overline{B} dv \qquad (2.19)$$

where $\overrightarrow{F}_{arm}$ -Force of armature, $\overrightarrow{J}$ –current density of the pulse generator, and $\overrightarrow{B}$ is de-magnetic flux density of the rails winding.

Obtaining the previously presented parameters at the highest possible values can only be done by superconducting coils. Figure 2.25 shows the operation of an electromagnetic launcher using an SMES. In the first phase, the superconducting coil is in the process of charging, contacts K1 being in the closed position and K2 and K3 in the open position. When the contact K3 closes and K1 and K2 remain in the open position, the coil is in the process of storage. To supply the launcher, the contact K3 opens and the contact K2 closes, and the launcher receives a pulse of energy.

B. Applications of SMES in renewable energy sources

The renewable sources like superconducting devices are used especially in storage energy. SMES unit stores energy for a long period of time in the magnetic field generated by DC flowing through a superconducting coil.

The electric power is stored in the magnetic field of a large superconducting magnet can be retrieved efficiently at short notice and offers a high efficiency up to 90% in the energy storing and releasing process.

Hence, the majority of commercial applications are based on LTS, which is a more mature technology.

Because of the capacity of quickly discharging or absorbing energy, SMES applications are mainly found in power networks for power quality improvement, low frequency damping, system stability control, and uninterruptible power supplies.

SMES systems have a great potential for supporting renewable systems within distributed generation networks; however, there is a lack of demonstrators which would provide a deeper insight into the long-term behavior of such systems. Subsidies and investment in such projects could provide researchers and companies with essential data sets that would enable them to investigate the technical, commercial, and regulatory aspects of SMES integration, which would lead to a better understanding and development of this technology.

Another renewable technology that would unlock the potential of SMES is the LH_2 fuel usage.

Fuel cells can be used very efficiently to replace conventional generation in both vehicles and power systems. The LH_2 storage tanks, however, can be used for keeping SMES coils under certain HTS and MTS transition temperatures, significantly decreasing the operating cost and creating the opportunity of a hybrid system [77, 78].

1. SMES with photovoltaic panels

Problems related to the depletion of fossil fuels, their rising prices as well as the issue of greenhouse gases, have led to the development of photovoltaic systems in terms of efficiency, production capacity, and lower production costs [64]. Utilizing more renewable energy sources (RES) minimizes the dependency on imported fossil fuels and creates sustainable energy production. However, RES is highly intermittent and fluctuating in nature. Particularly, the PV system is highly reliant on climatic or geometrical conditions and output power severely oscillates during bad climatic situations.

The higher fluctuations in the injected power will seriously affect the power system stability. Hence, the ESS is an essential element to mitigate these variations and improves the system power quality.

Numerous studies have shown that coupling SMES installations directly to wind or PV installations can vastly improve the voltage stability by either absorbing excessive power when there is extra generation or by discharging it in the grid if the voltage is suddenly dropping. In Figure 2.26, a method to connect an SMES with a PV's panel is presented [79].

2. SMES with wind farms

In the last years, the wind industry has expanded, the energy obtained from the wind being promising. For this reason, there has been an interest among energy companies. In 2009, the world's total generation capacity of wind power was 157.9 million kW, of which ~31% was newly installed within the year [16]. The brilliant future of wind power emphasizes the motivation on the technical upgrades in the wind farms, including the introduction of various HTS devices. In the past decade, many research and test operation efforts were paid on the new and high efficiency power applications, such as "direct- driven" permanent

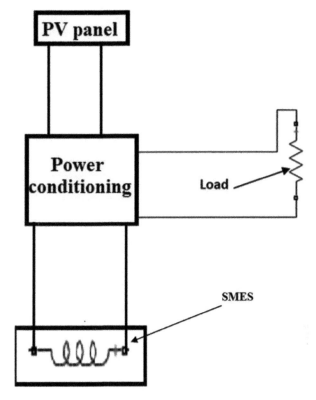

Figure 2.26   Application of SMES with PV panel

magnet (PM) generators and HTS generators; magnet, flying-wheel, and battery EESs; FCL; and solid-state transformers and electronic voltage regulators.
These devices are designed to solve the problems occurring in the quick boosting up of the wind farms and the strict requirements on connecting them to the
main frame of the power grids [80]. The problems that generally arise are related
to the optimization of the generator capacity, the size, and weight of the wind
turbine system, the stability of the output, as well as the tolerance of the system
to the fluctuations of the driving force, i.e. the wind and the load. One of the
key approaches to achieve the optimization is the superconducting technology.
Following this approach, a series of HTS devices were proposed, including HTS
generators, SMES, superconducting FCL, HTS transformers, and HTS power
transmission cables. In Figure 2.27, a method to connect an SMES with a wind
generator is presented [79].

3. Applications of SMES in hybrid energy storage systems

A hybrid energy storage system incorporates and uses the properties and quali-
ties of several storage devices, such as super capacitor or SMES batteries.

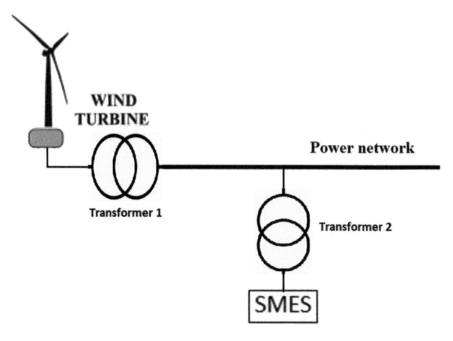

Figure 2.27   Application of SMES with Wind turbine adapted from Reference 79

By coupling a superconducting coil technology with high energy densities, such as batteries, compressed air storage, or pumped hydrostorage, a complete hybrid system can be created and used for a wider range of applications, including demand-side energy management with power quality control, peak shaving, and voltage stability [81]. Moreover, by forming hybrid systems, the cost can be significantly decreased: for the same energy storage capacity, a hybrid energy storage system with SMES and pumped hydroenergy storage can bring savings of over 90% in comparison with a sole SMES systems with the same capacity [82, 83]. A hybrid topologies is presented in Figure 2.28.

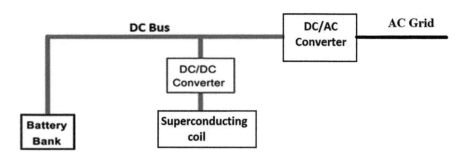

Figure 2.28   Hybrid storage topologies adapted from Reference 84

2.7 Practical aspects

In this sub-chapter, certain practical aspects regarding the design of the SMES are described. The influence of the manufacturing process over the electric, magnetic, mechanical, and thermal SMES parameters is discussed. The importance of the modeling and simulation and suggestions for stress reduction are included.

1. Design, testing, and computational aspects

Despite of the achievements in fundamental knowledge and engineering experience obtained by the evolution of the SMES technology, certain practical aspects still represent challenges [85–87]. Efficient management of heat flux flow in the superconducting coil that allows the best heat transfer and avoids quench, it is of great importance for the development of SMES systems. In literature, there are many presentation of designing SMES systems that store a certain amounts of energy [84, 88].

Modeling and simulation of the cables and superconducting coil can generate useful and significant results [23, 24]. A useful issue for modeling and simulation includes coupled analysis: magneto-structural-thermal analysis. Extensive simulations and computations should be performed on the SMES system design along with experimental verification and validation. It will be important to map out the 3D space of the associated magnetic field for particular designs resulting from the electric current in the SMES coils.

But the real system is affected by manufacturing process of the components as well as the winding process of the magnet. The fabrication process of the magnet could add another variations regarding electric, magnetic, mechanic, and thermal interactions in the configuration of the magnet. And all those perturbing factors should be well controlled in order to achieve the overall desired performance of the SMES system.

Furthermore, how the dynamic energy input and output conditions presented by real application affect the performance of the SMES magnet should be also learned. After all, a higher degree of optimization effort through extensive testing and design iterations will be essential to ensure the consistency and reliability of a matured technology.

A deeper understanding of the electric and magnetic effects considering another arrangements of the magnet components in order to reducing the stray field is still relatively less developed area but with possible improvements.

Complete testing at component and system levels is very useful, especially for SMES prototypes in an electrical grid in order to achieve their purposes. Testing must be conceived to allow a comparison with other systems that may achieve similar goals.

2. Improvement of the protection system

The energy stored in the superconducting magnet can be released in a very short time. The power per unit mass does not have a theoretical limit and can be extremely high (100 MW/kg), so the system must be contained in an excellent electric isolation. When a coil failure occurs the energy has to be released in order to avoid the coil damaging. Some conceptual designs of SMES system propose the absorption of energy by the superconducting cable and the support structure in case of system failure. But improvements still have to be studied.

3. Reducing the precooling time

Because the requested time for cooling the superconducting magnet from room temperature to operating temperature it is of months order, certain techniques to reduce it are necessary [25].

4. Reducing the stress

Because the superconducting coils are subjected to high values of Lorentz forces and stress generated by them, certain solutions for reinforcement structure to reduce them have to be found.

There are some high-strength composite materials that offer interesting perspectives for the future because their stress density ratio is very high. High-strength aluminum alloys are also excellent candidates: they have approximately 1/3 of the steel density. The mechanical design of an SMES is of extreme importance; the magnet conductor must be designed to withstand high stresses and deformations without degradation of superconducting properties [11, 26, 89, 90].

2.8   SMES cost analysis

In this sub-chapter, certain aspects regarding the costs of SMES systems are presented. The costs of the superconductive conductors, coolants (cryogens), and the refrigeration system are evaluated. Also a comparison among different types of storage systems regarding costs of the energy in $/kWh and power in $/kW will be presented.

2.8.1   Cost of the superconductors

The cost of an SMES depends mainly on the cost of the superconductive conductors and the refrigeration system.

The superconductor industry works along with the cryogenic industry because the operating temperatures are below 130K. Those low temperatures are obtained with cryogens or cryocoolers. Helium is the main cryogen with a boiling point at 4.2K or at 1.8 K for super fluid helium, and it is used for most low temperature magnet applications. But the price of helium is a problem. It costs between £7 and £9 per liter whereas LN is just £ 0.5 per liter [24, 62].

Operating at temperatures around 77 K is a great benefit in energy usage, efficiency, refrigeration reliability, and cost. One of the main drawbacks of the SMES is the cost of the superconducting wire that represents the majority of the capital cost. Electrical wires are generally compared on a per kiloampere meter (kAm) basis, where the kAm refers to the operating current level.

For HTS wires, copper is the main competitor for many of the electric power applications, being a reference and a target for the comparison with other electrical wires, the price of copper wire being about $24–36/kAm [24, 49].

HTSs in operating range 20–77 K could be economical for some applications considering the price of \$10–100/kAm. Generally, in the literature, the superconductors target price is around of \$25/kAm in the vicinity of the copper price. The LTS, Nb-Ti, and Nb3Sn are the cheapest superconducting wires available, and it has a price of around just \$1–2/kAm, and Nb3Sn has a cost of around \$11/kAm. These LTS superconductors require operation below 4.2K with liquid helium and cryocoolers, and for large-scale operations they become prohibitive.

For 1G cuprate HTS superconductors that begin to be commercialized, the product price is still the main obstacle for their large-scale application.

The current price is about \$60–80/kAm for Bi-2212 and Bi-2223 and \$100–200/kA m for REBCO conductors for use at 4.2 K and 10 T. Decreasing the price is the main aim for researchers and companies [24, 48, 49].

The MTS MgB2 operates at a temperature too low to be used with LN even though cost projections predict MgB2 be just a few \$/kAm in the near future.

The 2G HTS wire costs around \$300–400/kAm, while for 1G HTS the costs are around \$140–180/kAm, but they are more difficult to manufacture in long lengths. So far, the high cost of HTS materials has inhibited commercial adoption of HTS power devices. Estimates of the maximum acceptable price for different applications range from \$1 to \$100 per kAm.

Many researches have to be involved in superconducting materials improvement in order to simplify the manufacturing processes and finding new superconducting materials that could reduce the cost of SMES with at least 30%.

SMES is an emerging technology with many challenges that need to be surpassed in order to be competitive in the energy storage market. The superconductor along with the associated refrigeration requirement has always been an issue for SMES and will continue to be in the near future.

The cost of refrigeration using LN is much lower compared with liquid helium, allowing it to be used for large-scale industrial applications based on SMES. Early commercial projects using 2G HTSs are starting to appear but only on special applications because of the high price about \$300/kAm and the competition from super capacitors and flywheels [49].

2.8.2 Cost of the stored energy

In References 18, 24, 46, the main storage technologies are compared regarding to the ranges of values for power density, energy density, and costs. It can be noticed that the specific energy density of the SMES technology varies from 1 to 10 Wh/kg and the power density from 500 up to 2000 W/kg, being one of the main advantages of this technology. The cost for power density is reasonable, between 200 and 500 \$/kW being in the neighborhood of the super capacitors and of the flywheels. The price cost per unit energy is high, close to the super capacitors.

The price of SMES can be analyzed considering the superconducting coil configuration, solenoid/toroid, and the amount of the stored energy.

In Reference 25, a preliminary study of SMES system design and cost analysis for power grid application is presented. Both solenoid and toroid configurations, made of SCS121000 YBCO tape, are designed and compared.

For a small capacity superconducting magnet, the solenoid type geometry will be the first choice. The simulation results show that a higher radius helps to reduce the maximum magnetic field and increase the storage capacity. However, more superconductors are required with the increase in the coil radius. Applying a larger current into conductor is an efficient way to increase the stored energy. However, the maximum possible current is limited by the magnetic field produced by the current. Therefore, it is necessary to find an equilibrium between increasing energy capacity and ensuring stability.

The capacities of SMES systems are designed on different scales including kJ-, MJ-, and GJ-class for different applications. Finally, the costs are calculated based on the commercial price of superconductors.

Two solenoidal and two toroidal coils are considered storing 1.2 kJ, 1.6 MJ, 1.3 MJ, and 1GJ, respectively. They have the corresponding total length of the tapes 0.17, 12, 5, and 533 km, respectively. The cost of a 12 mm-width YBCO tape was considered 85 \$/m [24, 25].

The costs for kJ, MJ (solenoid and toroid configuration), and GJ are \$14,000, \$1,020,000, \$425,000, and \$450 million, respectively. With the energy increasing, the unit price of SMES systems decrease significantly, having the following price in \$/J: 1.4, 0.64, 0.33, and 0.045, respectively. The result indicates that the solenoid configuration is suitable for kJ-scale SMES, while the toroid configuration is suitable for GJ-scale SMES.

A full economic analysis comprises the cost estimation life cycle, which includes the capital cost, the operating cost, and the maintenance cost. Concerning the capital cost, two main parts need to be considered: energy capacity-related cost and power conversion-related cost. The former consists of capital and construction costs of superconductors, magnet structure components, cryogenic vessels, and refrigeration, protection, and control circuits. The latter is the cost of a required power electronics circuit. Different practical applications require different power electronics circuits and the costs vary for each situation.

2.9 Chapter summary

The central topic of this chapter is the presentation of energy storage technology using superconducting magnets. For the beginning, the concept of SMES is defined in 2.2, followed by the presentation of the component elements, as well as the types of geometries used in 2.3. Aspects of mechanical nature due to the Lorentz force occurring inside the superconducting coils are of particular importance in the proper functioning of the superconducting devices. For these reasons, these aspects have been carefully treated and accompanied by case studies for the calculation of the forces produced by the magnetic field for cylindrical and toroidal geometries. Section 2.3.3 presents a study of the calculation of forces produced by the magnetic field inside the cylindrical and toroidal superconducting coils. A case study on this topic is also described. The following section 2.4 contains elements of SMES dynamics, i.e. different methods of connecting an SMES to the network for different charge–discharge cycles. A numerical study case performed in Simulink® is presented. Section 2.5 deals with issues related to the nature of the materials from which the superconducting devices are made and also with the main cooling methods. Next, in 2.6 the material contains various applications of SMES such as storing energy from renewable sources, improving the parameters of transmission lines, electromagnetic launchers, superconducting cables, transformers, etc. In section 2.7 practical applications of SMES are presented. At the end of the chapter, in 2.8, a cost analysis of these devices is presented.

References

[1] Masih P. *Interplay between magnetism and superconductivity in high Tc superconductors* [online]. Available from https://shodhganga.inflibnet.ac.in/handle/10603/6620 [Accessed 28 Oct 2021].

[2] Farhadi M., Mohammed O. 'Energy storage technologies for high-power applications'. *IEEE Transactions on Industry Applications*. 2016, vol. 52(3), pp. 1953–61.

[3] *Five steps to energy storage world energy council 2020* [online]. Available from www.worldenergy.org/experiences-events/past-events/entry/five-steps-to-energy-storage-business-models-and-technologies [Accessed 13 Aug 2021].

[4] Koohi-Fayegh S., Rosen M.A. 'A review of energy storage types, applications and recent developments'. *Journal of Energy Storage*. 2020, vol. 27, p. 101047.

[5] Ayesha K. 'Green nanocomposites for energy storage'. *Journal of Composites Science*. 2021, vol. 5(201).

[6] Ciceron J. 'High energy density superconducting magnetic energy storage with second generation high temperature superconductors' 20 mars'.n.d.

[7] Zimmermann A.W., Sharkh S.M. 'Design of a 1 MJ/100 kw high temperature superconducting magnet for energy storage'. *Energy Reports*. 2020, vol. 6, pp. 180–88.

[8] Rong C., Barnes P. 'Developmental challenges of SMES technology for applications'. *IOP Conference Series Materials Science and Engineering*. 2017, vol. 279.

[9] Alattar A.H., Selem S.I., Metwally H.M.B, *et al.* 'Performance enhancement of micro grid system with SMES storage system based on mine blast optimization algorithm'. *Energies*. 2017, vol. 12(16), p. 3110.

[10] Badel A. 'Superconducting magnetic energy storage haute température critique comme source impulsionnelle' in *Supraconductivité [cond-mat.supr-con]. institut national polytechnique de grenoble-INPG*. Institut National Polytechnique de Grenoble - INPG; 2010.

[11] Tixador P. 'Superconducting magnetic energy storage: status and perspective' in *IEEE/CSC & esas european superconductivity news forum*;

[12] Ciceron J., Badel A., Tixador P., Razek A. 'Superconducting magnetic energy storage and superconducting self-supplied electromagnetic launcher'. *European Physical Journal Applied Physics*. 2017, vol. 80(2), p. 20901.

[13] Vadan I., Cziker A.C. *"Sistememoderne de Conversie Aenergiei" Editurau.T. Pres Cluj-Napoca'*. 2017.

[14] Tsutsui H., Nomura S., Watanabe N., *et al.* 'Distribution of stress in force-balanced coils on virial theorem'. *IEEE Transactions on Appiled Superconductivity*. 2003, vol. 13(2), pp. 1840–43.

[15] Tsutsui H., Kajita S., Ohata Y., Nomura S., Tsuji-Iio S., Shimada R. 'FEM analysis of stress distribution in force-balanced coils'. *IEEE Transactions on Appiled Superconductivity*. 2017, vol. 14(2), pp. 750–53.

[16] Rembeczki S. *Design and optimization of force-reduced high field magnets*. Hungary: Master of Science in Physics University of Debrecen; 2003.

[17] Tsutsui H., Habuchi T., Tsuji-Iio S., Shimada R. 'Analysis of stress distribution in helical coils with geodesic windings based on virial theorem'. *IEEE Transactions on Applied Superconductivity*. 2012, vol. 22(3), pp. 4705604–4705604.

[18] Zhu J., Zhang H., Yuan W., Zhang M., Lai X. 'Design and cost estimation of superconducting magnetic energy storage (SMES) systems for power grids'. *2013 IEEE Power & Energy Society General Meeting*; Vancouver, BC, 2013. pp. 1–5.

[19] Sangyeop K., Seyeon L., Sangyeop L, *et al.* 'Design of HTS magnets for a 2.5 MJ SMES'. *IEEE Transactions on Applied Superconductivity*. 2009, vol. 19(3), pp. 1985–88.

[20] Causley R., David C., Gower S. *'Design of a High Temperature Superconductor Magnetic Energy Storage Systems.'* 2001, pp. 322–25.

[21] Iwasa W. *Case studies in superconducting magnets*. US; 2009.

[22] Hassenzahl W.V., Hazelton D.W., Johnson B.K., Komarek P., Noe M., Reis C.T. 'Electric power applications of superconductivity'. Proceedings of the IEEE. 2004, vol. 92(10), pp. 1655–74.

[23] Ciceron J., Badel A., Tixador P., Forest F. 'Design considerations for high-energy density SMES'. *IEEE Transactions on Applied Superconductivity*. 2015, vol. 27(4), pp. 1–5.

[24] Shaw C. *Superconducting magnetic energy storage system commercialisation and marketing challenges* [Final Report]

[25] Available from https://www.semanticscholar.org/paper/Superconducting-Magnetic-Energy-Storage-System-%E2%80%93-Shaw/3c9b1f22b0d06b0c2d9ecbcd88f9b3b085d7f2eb

[26] Zhu J., Zhang H., Yuanb W., Zhang M., Lai X. 'Design and cost estimation of superconducting magnetic energy storage (SMES) systems for power grids' in *978-1-4799-1303-9/13/ $ 31.00 ©2013 IEEE*;

[27] *Magnet cutaway [online]*. Available from http://nmr.chem.umn.edu/cutaway.html

[28]   'Toral F "Mechanical design of superconducting accelerator magnets'. Presented at Conference: CAS - CERN Accelerator School; Erice (Italy).

[29]   Morandi A., Fabbri M., Gholizad B., Grilli F., Sirois F., Zermeno V.M.R. 'Design and comparison of a 1-MW/5-s HTS SMES with toroidal and solenoidal geometry'. *IEEE Transactions on Applied Superconductivity*. 2016, vol. 26(4), pp. 1–6.

[30]   Zimmermann A., Review of the State of the Art Superconducting Magnetic Energy Storage Presented at 5th International Conference on Applied Reaserch in Electrical, Mechanical & Mechatronics Engineering; 2019.

[31]   Zimmermann A.W., Sharkh S.M. 'Design of a 1 mj/100 kw high temperature superconducting magnet for energy storage'. *Energy Reports*. 2020, vol. 6, pp. 180–88.

[32]   Rong C., Barnes P. 'Developmental challenges of SMES technology for applications'. *Materials Science and Engineering*. 2017, vol. 279(1), p. 012013.

[33]   Wen J., Jin J.X., Guo Y.G., Jian G. *theory and application of superconducting magnetic energy storage* [online]. Available from http://citeseerx.ist.psu.edu/viewdoc/download/doi=10.1.1.554.3242&rep=rep1&type=pdf [Accessed 12 Dec 2021].

[34]   Available from http://www.phys.ubbcluj.ro/~iosif.deac/courses/THC/TH_C11.pdf

[35]   Chen X.Y., Feng J., Tang M.G., Xu Q., Li G.H. 'Superconducting magnetic energy exchange modelling and simulations under power swell/sag conditions'. *Energy Procedia*. 2017, vol. 105, pp. 4116–21.

[36]   Jin J.X., Chen X.Y., Qu R., Xin Y. 'A superconducting magnetic energy exchange model based on circuit-field-superconductor-coupled method'. *IEEE Transactions on Applied Superconductivity*. 2015, vol. 25(3), pp. 1–6.

[37]   Ortega A., Milano F. 'Generalized model of VSC-based energy storage systems for transient stability analysis'. *IEEE Transactions on Power Systems*. 2015, vol. 5, pp. 3369–80.

[38]   Farhadi M., Mohammed O. 'Energy storage technologies for high-power applications'. *IEEE Transactions on Industry Applications*. 2017, vol. 52(3), pp. 1953–61.

[39]   Ali Mohd H., Wu B., Dougal R.A. 'An overview of SMES applications in power and energy systems'. *IEEE Transactions on Sustainable Energy*. 2010, vol. 1(1), pp. 38–47.

[40]   González W., Garcés A., Escobar A. 'A generalized model and control for supermagnetic and supercapacitor energystorage'. *Ingeniería y Ciencia*. 2017, vol. 13(26), pp. 147–71.

[41]   Kumar A., Mohanty N., Anupriya M. 'Modeling and simulation of superconducting magnetic energy storage systems'. *International Journal of Power Electronics and Drive Systems (IJPEDS)*. 2015, vol. 6(3). Available from http://iaescore.com/journals/index.php/IJPEDS/issue/view/286

[42]   Ali Mohd.H., Wu B., Dougal R.A. 'An overview of SMES applications in power and energy systems'. *IEEE Transactions on Sustainable Energy*. 2010, vol. 1(1), pp. 38–47.

[43]   Shi J., Tang Y., Yang K., *et al.* 'SMES based dynamic voltage restorer for voltage fluctuations compensation". *Appl. Supercond.* 2010, vol. 20(3), pp. 1360–64.

[44]   Sahoo A.K., Mohanty N., M A. 'Modeling and simulation of superconducting magnetic energy storage systems'. *International Journal of Power Electronics and Drive Systems (IJPEDS)*. 2015, vol. 6(3), p. 524. Available from http://iaescore.com/journals/index.php/IJPEDS/issue/view/286

[45]   Sahoo A.K., Mohanty N., M A. 'Modeling and simulation of superconducting magnetic energy storage systems'. *International Journal of Power Electronics and Drive Systems (IJPEDS)*. 2015, vol. 6(3), p. 524. Available from http://iaescore.com/journals/index.php/IJPEDS/issue/view/286

[46]   Brito A. *Dynamic modelling.* Croatia: INTECH; 2010 Jan. p. 290.

[47]   Evans L. 'The large hadron collider: a marvel of technology, © 2019, open access EPFL press'.n.d.

[48]   Yao C., Ma Y. 'Superconducting materials: challenges and opportunities for large-scale applications'. *IScience.* 2021, vol. 24(6), p. 102541.

[49]   Kalsi S.S. *Applications of high temperature superconductors to electric power equipment* [online]. Hoboken, NJ; 2011 Mar 28. Available from http://doi.wiley.com/10.1002/9780470877890

[50]   Bussmann-Holder A., Keller H. 'High-temperature superconductors: underlying physics and applications [online]'. *Zeitschrift Für Naturforschung B.* 2020, vol. 75(1–2), pp. 3–14. Available from https://doi.org/10.1515/znb-2019-0103

[51]   Escamez G. 'Ph.D.thesis AC losses in superconductors:a multi-scale approach for the design of high current cables'.University Grenoble Alpes

[52]   Bray J.W. 'Superconductors in applications; some practical aspects'. *IEEE Transactions on Applied Superconductivity.* 2020, vol. 19(3), pp. 2533–39.

[53]   Senatore C., Alessandrini M., Lucarelli A., Tediosi R., Uglietti D., Iwasa Y. 'Progresses and challenges in the development of high-field solenoidal magnets based on RE123 coated conductors'. *Superconductor Science & Technology.* 2014, vol. 27(10), pp. 103001–2014.

[54]   Han P., Wu Y., Liu H., Li L., Yang H. 'Structural design and analysis of a 150 kj HTS SMES cryogenic system'. *Physics Procedia.* 2015, vol. 67, pp. 360–66.

[55]   Drozdov A.P., Eremets M.I., Troyan I.A., Ksenofontov V., Shylin S.I. 'Conventional superconductivity at 203 kelvin at high pressures in the sulfur hydride system'. *Nature.* 2015, vol. 525(7567), pp. 73–76.

[56]   Snider E., Dasenbrock-Gammon N., McBride R., *et al.* 'Room-temperature superconductivity in a carbonaceous sulfur hydride'. *Nature.* 2020, vol. 586(7829), pp. 373–77.

[57]   Wilson W. *Superconducting magnets.* Oxford, UK: Clarendon Press; 1987.

[58]   Breschi M., Trevisani L., BotturaL. D.A., Trillaud F. 'Comparing the thermal stability of NBTI and NB3SN wires' supercond'. *Sci. Technol.* 2009, vol. 22, p. 025019.

[59] Radenbaugh R. 'Refrigeration for superconductors'. *Proceedings of the IEEE*. 2004, vol. 92(10), pp. 1719–34.

[60] Marquardt E.D., Radebaugh R., Doba J. 'A cryogenic catheter for treating heart arrhythmia'. [Plenum Press, NY] *Journal of Advances in Cryogenic Engineering*. 1998, vol. 43, pp. 903–10.

[61] Ross R.G.J. *refrigeration systems for achiving cryogenic temperatures* [online]. Available from https://doi.org/10.1201/9781315371962

[62] Green M.A. 'The cost of coolers for cooling superconducting devices at temperatures at 4.2 K, 20 K, 40 K and 77 K'. *IOP Conference Series*. 2015, vol. 101, p. 012001.

[63] Green M.A. *Estimating the cost of superconducting magnets and the refrigerators needed to keep them cold*. Berkeley, CA: Lawrence Berkeley National Laboratory;

[64] Xu Chu C., Xiaohua J., Yongchuan L., Xuezhi W., Wei L. 'SMES control algorithms for improving customer power quality'. *IEEE Transactions on Appiled Superconductivity*. 2001, vol. 11(1), pp. 1769–72.

[65] Zhu J., Zhang H., Yuan W., Zhang M., Lai X. 'Design and cost estimation of superconducting magnetic energy storage (SMES) systems for power grids'. *National Natural Science Foundation of China*. 2013.

[66] Nuno A*et al.* 'Integration of SMES devices in power systems-opportunities and challenges'. *2015 9th International Conference on Compatibility and Power Electronics (CPE)*; 2015. pp. 482–87.

[67] Xue X.D., Cheng K.W.E., Darmawan S. 'Power system applications of superconducting magnetic energy storage systems'. Presented at Conference Record - IAS Annual Meeting (IEEE Industry Applications Society);

[68] Buckles W., Hassenzahl W.V. 'Superconducting magnetic energy storage'. *IEEE Power Engineering Review*. 2000, vol. 20(5), pp. 16–20.

[69] Mishra M. 'Power quality disturbance detection and classification using signal processing and soft computing techniques: a comprehensive review'. *International Transactions on Electrical Energy Systems*. 2019, vol. 29(8). Available from https://onlinelibrary.wiley.com/toc/20507038/29/8

[70] Ribeiro P.F., Arsoy A., Liu Y. L. 'Transmission power quality benefits realized by a SMES-FACTS controller'. *9th International Conference on Harmonics and Quality of Power*; 2000. pp. 307–12.

[71] Jin J.X., Wang Q.L, *et al.* 'Enabling high-temperature superconducting technologies toward practical applications'. *IEEE Transactions on Applied Superconductivity*. 2014, vol. 24(5), pp. 1–12.

[72] Torre W. V., Eckroad S. 'Improving power delivery through the application of superconducting magnetic energy storage (SMES)'. *IEEE Power Engineering Society Winter Meeting Conference*; 2001. pp. 81–87.

[73] Lamoree J., Tang L., DeWinkel C., Vinett P. 'Description of a micro-SMES system for protection of critical customer facilities'. *IEEE Transactions on Power Delivery*. 1994, vol. 9(2), pp. 984–91.

[74] Kumar V., Savardekaro U. C. 'Superconducting transformer'. *Journal of New Innovations in Engineering and Technology*. pp. 67–73. n.d.

[75]   Kroczek R., Domin J. 'A COMSOL multiphysic using for designing A hybrid electromagnetic launcher'. Presented at COMSOL Conference 2010; Paris,

[76]   Ciceron J., Badel A., Tixador P., Razek A. 'Superconducting magnetic energy storage and superconducting self-supplied electromagnetic launcher'. *The European Physical Journal Applied Physics*. 2017, vol. 80(2), p. 20901.

[77]   Ali M.H., Wu B., Dougal R.A. 'An overview of SMES applications in power and energy systems'. *IEEE Transactions on Sustainable Energy*. 2010, vol. 1(1), pp. 38–47.

[78]   Ackermann T. *Wind power in power systems*. Chichester: Wiley; 2005.

[79]   Nielsen K.E., Molinas M. *2010 IEEE international symposium on industrial electronics (ISIE 2010); bari, italy, norwegian university of science and technology department of electric power engineering*; 2010.

[80]   Available from https://www.mathworks.com/products/matlab-[online] [Accessed 17 Jun 2021].

[81]   Etxeberria A., Vechiu I., Camblong H., Vinassa J.-M. 'Comparison of three topologies and controls of a hybrid energy storage system for microgrids'. *Energy Convers Manag*. 2012, vol. 54(1), pp. 113–21.

[82]   Rong C., Barnes P. 'Developmental challenges of SMES technology for applications conf'. *Series: Materials Science And*. 2017, vol. Engineering 279, p. 012013.

[83]   Qiu M., Rao S., Zhu Z, *et al.* 'Mechanical properties of MJ-class toroidal magnet wound by composite HTS conductor'., pp. ASC2016–3LPo1H.n.d.

[84]   Jubleanu R., Cazacu D., Bizon N. 'Hybrid energy storage – a brief overview'. *13th International Conference on Electronics, Computers and Artificial Intelligence (ECAI)*; Pitesti, Romania, 2021. pp. 1–15.

[85]   Mukherjee P., Rao V. 'Design and development of high temperature superconducting magnetic energy storage for power applications - a review'. *Journal of Superconductivity and It's Applications*. 2019, vol. 563, pp. 67–73.

[86]   Sirois F., Grilli F. 'Potential and limits of numerical modelling for supporting the development of HTS devices'. *Superconductor Science and Technology*. 2015, vol. 28(4), p. 043002.

[87]   Kondratowicz-Kucewicz B., Janowski T., Kozak S., Kozak J., Wojtasiewicz G., Majka M. 'Bi-2223 HTS winding in toroidal configuration for SMES coil'. *Journal of Physics*. 2010, vol. 234(3), p. 032025.

[88]   Abdullah Al Zaman M. d, Ahmed S., Nusrath M. *An overview of superconducting magnetic energy storage (SMES) and its applications*. Vols. 11–12. Bangladesh: BUET –Dhaka; 2018 Jan.

[89]   Cazacu D., Jubleanu R., Bizon N., Monea C. 'Comparative numerical analysis of the stored magnetic energy in cylindrical and toroidal superconducting magnetic coils'. *11th International Conference on Electronics, Computers and Artificial Intelligence (ECAI)*; Pitesti, Romania, 2019. pp. 1–5.

[90] Jubleanu R., Cazacu D., Bizon N. 'Stress in cylindrical and toroidal superconducting coils'. *12th International Conference on Electronics,
Computers and Artificial Intelligence (ECAI)*; Bucharest, Romania, 2020.
pp. 1–6. Available from https://ieeexplore.ieee.org/xpl/mostRecentIssue.
jsp?punumber=9218228

Chapter 3

Superconducting magnetic energy storage control methods

Sagnika Ghosh[1] and Mohd Hasan Ali[2]

3.1 Introduction

The decarbonization of the electric power grid has laid a path for the integration of renewable resources to the existing grids. The renewable resources along with the energy storage system are a robust system for the loads. The intermittent nature of the renewable energy resources such as wind system and solar system makes the presence of energy storage system more useful. The other usefulness of the energy storage system is to keep the active grid power stable. Any disturbances that occur in the power grid have a detrimental effect on the customer loads.

Energy storage system also contributes to the power transmission and distribution systems to improve the conditions such as subsynchronous resonance [1, 2], power system stability [3, 4], load frequency control [5, 6], and voltage dynamic regulation [7], and at the distribution level, it helps with power oscillations [8–11]. There are different types of energy storage systems such as battery energy storage system (BESS), flywheel energy storage (FES) system, pumped hydroelectric storage (PHS) system, and superconducting magnetic energy storage (SMES) system. BESS has lower efficiency and shorter service life per load/discharge cycle than the SMES. The FES system requires more maintenance and has a slower response time than the SMES system. In PHS, limitations come from the topographic conditions and access to water. The intermittent nature of the renewable system is overcome by the energy storage systems. The integration of renewable resources also increases the oscillations in the power system. SMES deals with these issues arising in the power system [12–14]. SMES has quite a few advantages like a fast response, high energy storage efficiency, and large discharge within a short period of time. This makes SMES widely used in power distribution systems with high

[1]Tennessee State University, Nashville, TN, USA
[2]The University of Memphis, Memphis, TN, USA

penetration of renewable energy resources to improve the dynamic performances of the system [9].

SMES is a large superconducting coil that can store energy in the magnetic field generated by Direct Current (DC) flowing through it. The coil is cooled to a temperature below its critical temperature, thus reducing the losses during operation. SMES provides the highest densities of power storage [15]. It can charge and discharge very fast in a short period of time which provides a high dynamic response in the range of milliseconds. According to the power requirement, active and reactive power can be absorbed or released from the SMES coil. SMES coil also has a high life cycle. This means that the coils can withstand tens of thousands of charging cycles. The cooling of the SMES coil makes the efficiency a little lower, but the power required is far less than the output power of SMES. Due to limited losses, the efficiency of SMES system is about 90% [16].

Based on these backgrounds, this chapter deals with the basics of SMES device and its control techniques. Different control techniques for the voltage source converter (VSC) and current source converter (CSC) are discussed in detail.

3.2   Various control techniques of SMES

SMES systems require a power electronic converter to be integrated into the power distribution systems as well as fast and robust control strategies to guarantee their efficient performance [17, 18]. There are three types of configurations to integrate SMES systems, depicted in Figure 3.1, namely, line-commutated converters (LCC)

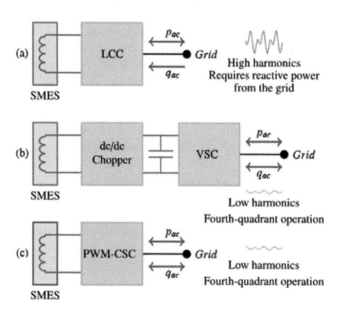

Figure 3.1   Interaction of SMES with grid; (a) connection using LCC, (b) connection using VSC, and (c) connection using CSC [11]

[9, 19], VSC with DC/DC-chopper connection [9, 20], and pulse-width modulated CSC (PWM-CSC) [8, 20]. The first type of converter uses thyristors as a switching device. This converter has low-switching losses and active power control; however, it has low capacity to control reactive power. In addition, an LCC requires passive filters to mitigate its harmonic injections. The VSC and PWM-CSC permit an independent control of active and reactive power in both directions with low-harmonic distortion.

To develop a control method, the VSC must control the line currents of Alternating current (AC) side with respect to the demand, and it should produce constant DC output voltage. In case of VSC, a DC/DC chopper to allow current variations on the superconducting coil is required, while a PWM-CSC directly allows this type of variations. For this reason, it is more natural to use a PWM-CSC to integrate an SMES system into an AC grid than other technologies [9, 21, 22]. The most classical controller proposed in the specialized literature to operate a power system conditioning using a SMES system is the proportional-integral (PI) strategy [23, 24]. Furthermore, this control methodology presents low performance and low robustness because the SMES system is a nonlinear and strongly coupled system [24, 25]. To design PI controller, Taylor's linearization of some partial linearization is employed, which implies that it is only possible to guarantee stability properties around the operational point [24]. Nevertheless, when there are important deviations concerning the operational point, the PI controller presents low performance, and it could have instability [24]. Some authors [26] have proposed an adaptable PI controller by adjusting its parameters via meta-heuristic optimization; notwithstanding, this approach requires more complex calculations and increases the difficulty of the controller. For the other part, nonlinear control methods such as fuzzy logic, sliding planes, or hysteresis techniques have been proposed to control the SMES system. These techniques of control have shown superior performance when compared with classical PI methods [27–30]. The fuzzy logic control can be applied to complex systems, enhancing the performance and robustness; nonetheless, it presents poor steady-state performance, and its implementation is highly dependent on the control problem. In the case of sliding methods, they present a high capacity to reject external perturbations and the possibility to operate with parameter variations; nevertheless, if there are vibration problems in the system, it will reduce the control performance. Finally, in the case of hysteresis control, it is easy to implement and has good performance under transient conditions; notwithstanding, the variable frequency of operation makes difficult to filter the design. On the other hand, it is also possible to find strategies such as linear control by state feedback [31, 32], model predictive controllers (MPCs) [18, 33], and passivity-based applications [34–36] among others.

3.2.1   VSC-based SMES

VSC-based control is used in high-power and high-performance applications because it provides constant DC-link voltage, bidirectional power flow, controllable power factor, and low-harmonic distortion. Between the VSC and the superconducting coil,

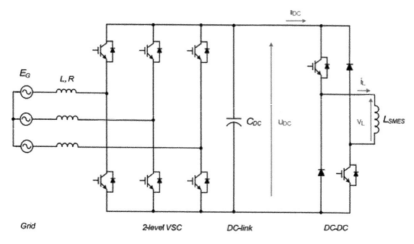

Figure 3.2   Two-level VSC [37]

a DC-DC converter is required to adapt the voltage levels of the DC link and the coil. VSC-based SMES allows independent control of reactive and real power flowing between the power system and the coil. VSC provides continuous rated capacity at low or no coil current [37]. There are two different types of VSC topologies: two-level VSC and three-level VSC. In the case of two-level VSC, it consists of two active insulated gate bipolar transistor (IGBT) switches with antiparallel diodes. The DC-DC converter is made up of IGBT with antiparallel diode and a diode as such in the Figure 3.2.

In three-level VSC, it consists of four active switches with four antiparallel diodes. This converter is a three-level neutral point clamped (NPC) converter as shown in Figure 3.3. The DC-link capacitor is divided into two, which provides a neutral point (NP). The diodes disconnected to the NP are the clamping diodes.

Figure 3.3 shows that the three-level NPC converter has some advantages over the two-level topology such as no dynamic voltage sharing problem. Each of the switches in the NPC converter withstands only half of the total DC voltage during commutation, static voltage equalization without using additional components. The static voltage equalization can be achieved when the leakage current of the top and bottom switches in a converter leg is selected to be lower than that of the inner switches, low total harmonic distortion (THD), and *dv/dt*. The waveform of the line-to-line voltages is composed of five voltage levels, which leads to lower THD and *dv/dt* in comparison to the two-level converter operating at the same voltage rating and device switching frequency.

The mathematical model in three-phase stationary frame is given as follows:

$$L\frac{di_a}{dt} = -Ri_a + u_{E_a} - \left(E_a - \frac{E_a + E_b + E_c}{3}\right)u_{dc}$$

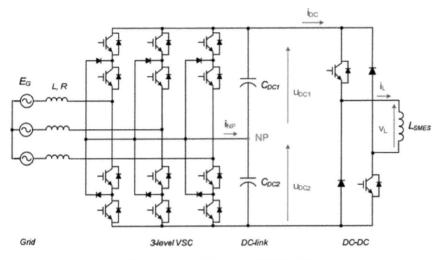

Figure 3.3   Three-level VSC [40]

$$L\frac{di_b}{dt} = -Ri_b + u_{E_b} - \left(E_b - \frac{E_a + E_b + E_c}{3}\right)u_{dc}$$

$$L\frac{di_c}{dt} = -Ri_c + u_{E_c} - \left(E_c - \frac{E_a + E_b + E_c}{3}\right)u_{dc}$$

$$C\frac{du_{dc}}{dt} = i_a E_a + i_b E_b + i_c E_c - i_{chopper}$$

The three-phase stationary frame model can be changed into synchronous rotating *dq* frame, and it is expressed as follows:

$$L\frac{di_d}{dt} = -Ri_d + \omega L i_q + u_{Ed} - u_{dc}$$

$$L\frac{di_q}{dt} = -Ri_q - \omega L i_d + u_{Eq} - u_{dc}$$

$$C\frac{du_{dc}}{dt} = \frac{3}{2}\left(i_q E_q + i_d E_d\right) - i_{chopper}$$

where ω is the angular frequency at the AC side, i_d and i_q are the current components in *d* axis and *q* axis, respectively, flowing through the *R* and *L* (transformer parameters). u_{Ed} and v_L depict the AC voltage of the main grid and the output AC voltage of the VSC, respectively. C_{DC} is the DC-link capacitor of the VSC, and the voltage across it is v_{DC} current delivered (or absorbed) by the SMES system, and Ls_{MES} is its inductance.

3.2.1.1 PI control method

The PI control method is most used in industry because of its robustness and wide range of stability.

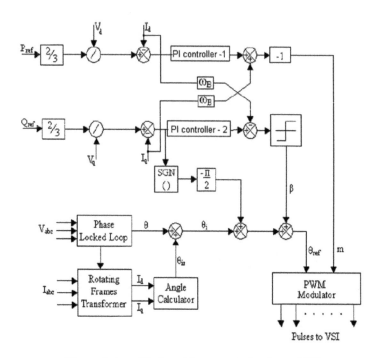

Figure 3.4   PI closed loop control scheme [38]

To use PI controller for SMES, the line current I_{abc} and the line voltage V_{abc} on the transmission line are converted into *d-q* components using Parks' transformation. The desired current references, namely I_{dref} and I_{qref}, are compared with actual current components I_d and I_q, respectively, and the error signals are processed in the PI controller [38]. Based on the controller parameters, the displacement angle is derived to control the voltage with respect to the line current.

Due to the nonlinearity of the system and sensitive parameter variations, different optimization techniques have been employed [39–41]. The PI controllers require complex analysis for fine tuning of the parameters. It takes a long time. This leads to the development of adaptive PI controllers. Figure 3.4 shows the PI closed loop control scheme.

3.2.1.2   Adaptable PI controller

Adaptive PI controller is used to control the VSC and the DC-DC converter to control the active and reactive power of the system [42]. The adaptive control provides a faster convergence and less computational complexity. The adaptive PI controllers, as shown in Figure 3.5, are used to generate sinusoidal waveform which are then compared with triangular waveform to generate IGBT's gate signals. The outer loop-adaptive controllers are used to correct the DC-link voltage (V_{DC}) error signal and the voltage at the point of common coupling error signal

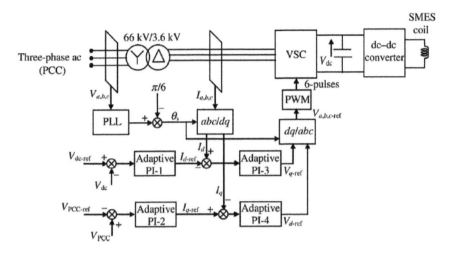

Figure 3.5 *Block diagram of the set-membership affine projection algorithm (SMAPA)-based adaptive PI controller [42]*

producing *d*-axes current and the *d*-axes reference $I_{d\text{-ref}}$ and $I_{q\text{-ref}}$, respectively. There are two other adaptive controllers used for inner loop to follow the I_d error signal and the I_q error signal producing $V_{q\text{-ref}}$ and $V_{d\text{-ref}}$, respectively.

3.2.1.3 Fuzzy logic control

In Reference 43, a fuzzy logic control method for the bridge type SMES has been discussed. In this control method, high temperature superconductors (HTSs) toroidal coil is charged or discharged by adjusting the average voltage across the coil. The average value could be a positive or a negative value. As the switches change their states, the coil is charged or discharged. The current is constant, while the HTS coil is on standby state. The current through the HTS coil is selected as inputs and the state of the HTS coil as output. Based on the defuzzification results, the microcontroller unit will multiply drive signals for the power metal oxide semiconductor field effect transistors and power transistors. Figure 3.6 shows the flowchart that the fuzzy logic controller follows.

3.2.1.4 Linear control by state feedback

State-feedback control is used when the state which could not always act correctly as, caused by faults (such as short circuit), the operating points are far away from the equilibrium points at which the approximate linearization models are formed. This controller is designed with inner loop and outer loop in a cascaded structure [44]. The state of the DC chopper is used to regulate the control of voltage on the SMES coil. This is achieved by active power balance between the AC and DC side so that the voltage control loop is treated as active current. The output of the current loop is determined by the reactive current. The VSC is controlled by the PWM control signal.

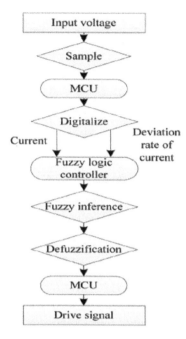

Figure 3.6   Step of fuzzy logic controller design [43]

3.2.1.5   Model predictive controller

In all the different types of control of SMES, a PWM is used which creates harmonics in DC current for the coil. Eddy current losses increase due to the second and third harmonics [45]. MPC has been widely used in VSC, multilevel converter, and matrix converter [44]. MPC reduces voltage and current ripples [33]. In Reference 46, MPC is applied to the converters of SMES system. The MPC is applied for both VSC converter and DC-DC chopper. For each possible voltage vector, this control model can predict the future value of load current suing a discrete time model

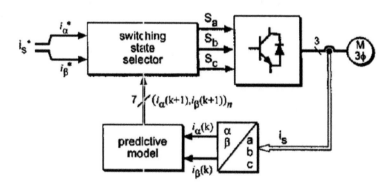

Figure 3.7   Predictive current control [44]

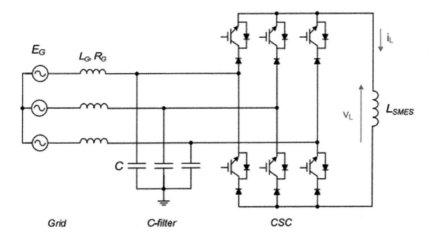

Figure 3.8   Two-level CSC [37]

[47]. The vector decreases the load current error, and it is applied during the next sampling interval. The switching state could be changed only once at each sampling instant. Predictive control method is useful for high-bandwidth applications. Figure 3.7 shows the model for the predictive controller.

3.2.2   CSC-based control

As shown in Figure 3.8, two-level CSC consists of two IGBTs and two blocking diodes. There are also capacitor banks used here to help with the commutation of the switching devices like the IGBTs. It also provides a clear path for the energy stored in the inductance of each phase. This acts as a harmonic filter, improving the load current and voltage; otherwise, a high-voltage spike would be induced, damaging the switching devices. By reversing the DC polarity, bidirectional power flow is achieved. The DC side of the CSC is connected to the coil directly, and the AC side is connected to the grid via C-filter (capacitor bank).

There are few nonlinear control strategies applied to SMES systems with PWM-CSC, which represents an opportunity for research. In the case of CSC, there are also control strategies as decoupled state-feedback control [48], classical PI control [S10], linear-quadratic regulator (LQR) control [49], power control theory [50], and fuzzy logic control [51] among others.

The mathematical model of CSC consists of two parts; one is the active and reactive current control, and the other is the current coil of the superconducting coil. The mathematical model in the synchronous rotating *d-q* frame can be shown as [52]

$$L\frac{di_d}{dt} = u_{sd} - u_d - i_d R + \omega L i_q$$

$$L\frac{di_q}{dt} = u_{sq} - u_q - i_q R + \omega L i_d$$

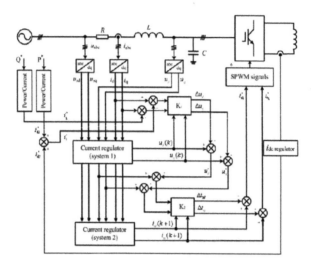

Figure 3.9   Decoupled state-feedback controller [53]

$$C\frac{du_d}{dt} = i_d - s_d i_{dc} + \omega C u_q$$

$$C\frac{du_q}{dt} = i_q - s_q i_{dc} + \omega C u_d$$

where ω is the angular frequency at the AC side, and i_d and i_q are the current components in d and q axes, respectively [53].

3.2.2.1   Decoupled state-feedback control

In this type of CSC control, two control loops are considered as shown in Figure 3.9. One loop deals with the active and reactive current, which realizes the decoupled control on active and reactive power transfer of SMES, and the other is for controlling the current of the superconducting coil, which maintains the balance between the AC and the DC side of power-conditioning system [54]. State-feedback control also reduces the THD. In this control design technique, the effect of control time delay and the current error compensation is considered to effectively transfer power. The output voltage error vector is added as the additional state variables, and a linear state-feedback loop is added to the control system to achieve the feature of deadbeat.

3.2.2.2   LQR control

The design of the regulator is to determine the optimal control law which can transfer the system from its initial state to the final state such that a given performance index is minimized [55]. The performance index that is widely used in optimal control design is known as the quadratic performance index. This is based on minimum error and minimum energy criteria. The LQR control method significantly improves peak deviations and settling time.

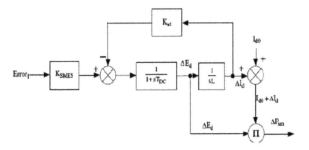

Figure 3.10 *Block diagram of SMES with negative inductor current deviation feedback used in the model of two-area system with LQR-SMES [54]*

Figure 3.10 shows the block diagram of SMES with negative inductor current deviation feedback. Due to a sudden change in loading, the power needs to be pumped back into the grid, the control voltage, E_d is negative as the current through the inductor and the thyristors cannot change its direction.

3.2.2.3 Power control theory

The power control theory improves the stability of the system. It is a control of active power source, can participate in dynamic behavior, and control the damping torque and the synchronization torque. The power-conditioning system of SMES is controlled by outer- and inner-loop control [56]. The converter control generates the triggering pulse for switching devices.

3.2.2.4 Nonlinear feedback

A nonlinear feedback analysis to design a control strategy for an SMES system is proposed based on the expected behavior of the superconducting coil device. It is also presented an analysis of stability under steady-state conditions by using time-domain reference frame via Lyapunov theory. Time domain simulations must demonstrate the soundness and adequate performance of the proposed controller in different cases. These cases contemplate unbalance in voltage, harmonic distortion in voltage, and power fluctuations caused by the high variation of weather resources.

3.3 Conclusion

This chapter deals with some basics of SMES and its control methodology. SMES is one of the most developing and efficient energy storage devices. The integration of SMES systems in the AC power microgrids under connected operation mode allows compensating active and reactive power dynamically, which clearly improves the grid performance in terms of power factor reducing at the same time the power oscillations produced by renewable generators. Different topologies of the VSC and CSC systems are explored. The different control methodologies for VSC and CSC are used to mitigate the variation in voltage and power for grid-connected systems using SMES.

References

[1]   Farahani M. 'A new control strategy of SMES for mitigating subsynchronous oscillations'. *Physica C*. 2012, vol. 483, pp. 34–39.

[2]   Gil-González W., Montoya O.D., Garces A. 'Control of a SMES for mitigating subsynchronous oscillations in power systems: a PBC-PI approach'. *Journal of Energy Storage*. 2018, vol. 20, pp. 163–72.

[3]   Ortega A., Milano F. 'Generalized model of VSC-based energy storage systems for transient stability analysis'. *IEEE Transactions on Power Systems*. 2018, vol. 31(5), pp. 3369–80.

[4]   Ali M.H., Murata T., Tamura J. 'Transient stability enhancement by fuzzy logic-controlled SMES considering coordination with optimal reclosing of circuit breakers'. *IEEE Transactions on Power Systems*. 2018, vol. 23(2), pp. 631–40.

[5]   Farahani M., Ganjefar S. 'Solving LFC problem in an interconnected power system using superconducting magnetic energy storage'. *Physica C*. 2018, vol. 487, pp. 60–66.

[6]   Shayeghi H., Jalili A., Shayanfar H.A. 'A robust mixed H2/H∞ based LFC of A deregulated power system including SMES'. *Energy Convers Manag*. 2008, vol. 49(10), pp. 2656–68.

[7]   Xiaohua H., Guomin Z., Liye X. '*IEEE Transactions on Applied Superconductivity*'.2010, vol. 20(3), pp. 1316–19.

[8]   Shi J., Tang Y., Ren L., Li J., Cheng S. 'Discretization-based decoupled state-feedback control for current source power conditioning system of SMES'. *IEEE Transactions on Power Delivery*. 2008, vol. 23(4), pp. 2097–104.

[9]   Ali Mohd.H., Wu B., Dougal R.A. 'An overview of SMES applications in power and energy systems'. *IEEE Transactions on Sustainable Energy*. 2008, vol. 1(1), pp. 38–47.

[10] Ibrahim H., Ilinca A., Perron J. 'Energy storage systems—characteristics and comparisons'. *Renewable and Sustainable Energy Reviews*. 2008, vol. 12(5), pp. 1221–50.

[11] Gil-González W., Montoya O.D. 'Passivity-based PI control of a SMES system to support power in electrical grids: a bilinear approach'. *Journal of Energy Storage*. 2018, vol. 18, pp. 459–66.

[12] Ngamroo I., Karaipoom T. 'Improving low-voltage ride-through performance and alleviating power fluctuation of DFIG wind turbine in DC microgrid by optimal SMES with fault current limiting function'. *IEEE Transactions on Applied Superconductivity*. 2014, vol. 24(5), pp. 1–5.

[13] Molina M.G., Mercado P.E. 'Power flow stabilization and control of microgrid with wind generation by superconducting magnetic energy storage'. *IEEE Transactions on Power Electronics*. 2018, vol. 26(3), pp. 910–22.

[14]   Kim A.-R., Kim G.-H., Heo S., Park M., Yu I.-K., Kim H.-M. 'SMES applica-
       tion for frequency control during islanded microgrid operation'. *Physica C.*
       2018, vol. 484, pp. 282–86.
[15]   Vazquez S., Lukic S.M., Galvan E., Franquelo L.G., Carrasco J.M. 'Energy
       storage systems for transport and grid applications'. *IEEE Transactions on
       Industrial Electronics.* 2018, vol. 57(12), pp. 3881–95.
[16]   Boom R.W., Peterson H. 'Superconductive energy storage for electric utili-
       ties—A review of the 20 year wisconsin program'. *Proceedings of the 34th
       International Power Sources Symposium*; Cherry Hill, NJ, 1990.
[17]   Montoya O.D., Garcés A., Serra F.M. 'DERs integration in microgrids us-
       ing vscs via proportional feedback linearization control: supercapaci-
       tors and distributed generators'. *Journal of Energy Storage.* 2018, vol. 16,
       pp. 250–58.
[18]   Montoya O.D., Garcés A., Espinosa-Pérez G. 'A generalized passivity-based
       control approach for power compensation in distribution systems using elec-
       trical energy storage systems'. *Journal of Energy Storage.* 2018, vol. 16, pp.
       259–68.
[19]   Farahani M., Ganjefar S. 'Solving LFC problem in an interconnected power
       system using superconducting magnetic energy storage'. *Physica C.* 2013,
       vol. 487, pp. 60–66.
[20]   Montoya O.D., Gil-Gonzalez W., Garces A. 'Control for EESS in three-
       phase microgrids under time-domain reference frame via PBC theo-
       ry'. *IEEE Transactions on Circuits and Systems II.* 2007, vol. 66(12),
       pp. 2007–11.
[21]   Giraldo E., Garces A. 'An adaptive control strategy for a wind energy conver-
       sion system based on PWM-CSC and PMSG'. *IEEE Transactions on Power
       Systems.* 2014, vol. 29(3), pp. 1446–53.
[22]   Montoya O.D., Gil-González W., Garcés A., Espinosa-Pérez G. 'Indirect
       IDA-PBC for active and reactive power support in distribution networks us-
       ing SMES systems with PWM-CSC'. *Journal of Energy Storage.* 2018, vol.
       17, pp. 261–71.
[23]   Ortega A., Milano F. 'Comparison of different control strategies for energy
       storage devices'. *2016 Power Systems Computation Conference (PSCC)*;
       Genoa, Italy, IEEE, 2016. pp. 1–7.
[24]   Liu F., Mei S., Xia D., Ma Y., Jiang X., Lu Q. 'Experimental evaluation of
       nonlinear robust control for smes to improve the transient stability of power
       systems'. *IEEE Transactions on Energy Conversion.* 2004, vol. 19(4), pp.
       774–82.
[25]   Tan Y.L., Wang Y. 'Stability enhancement using SMES and robust nonlinear
       control'. *International Conference on the Proceedings of Energy Management
       and Power Delivery*; IEEE, 1998. pp. 171–76.
[26]   Vachirasricirikul S., Ngamroo I. 'Improved H2=H1 control-based robust PI
       controller design of SMES for suppression of power fluctuation in microgrid'
       in *International electrical engineering congress (IEECON)*. IEEE; 2014. pp.
       1–4.

[27]   Lu Q., Sun Y., Mei S. *Nonlinear control systems and power system dynamics*. Vol. 10. Germany: Springer Science & Business Media; 2013.

[28]   Yi H., Zhuo F., Wang F., Wang Z. 'A digital hysteresis current controller for three-level neural-point-clamped inverter with mixed-levels and prediction-based sampling'. *IEEE Transactions on Power Electronics*. 2016, vol. 31(5), pp. 3945–57.

[29]   Flores-Bahamonde F., Valderrama-Blavi H., Bosque-Moncusi J.M., García G., Martínez-Salamero L. 'Using the sliding-mode control approach for analysis and design of the boost inverter'. *IET Power Electronics*. 2016, vol. 9(8), pp. 1625–34.

[30]   Tao C.W., Wang C.M., Chang C.W. 'A design of A DC–AC inverter using A modified ZVS-PWM auxiliary commutation pole and A DSP-based PID-like fuzzy control'. *IEEE Transactions on Industrial Electronics*. 2016, vol. 63, pp. 397–405.

[31]   Gil-Gonzalez W., Montoya O.D., Garces A., Escobar-Mejia A. 'Supervisory LMI-based state-feedback control for current source power conditioning of SMES'. *Ninth Annual IEEE Green Technologies Conference (GreenTech)*; Denver, CO, 2017. pp. 145–50.

[32]   Kiaei I., Lotfifard S. 'Tube-based model predictive control of energy storage systems for enhancing transient stability of power systems'. *IEEE Transactions on Smart Grid*. 2018, vol. 9(6), pp. 6438–47.

[33]   Nguyen T.T., Yoo H.J., Kim H.M. 'Applying model predictive control to SMES system in microgrids for eddy current losses reduction'. *IEEE Transactions on Applied Superconductivity*. 2016, vol. 26(4), pp. 1–5.

[34]   Hou R., Song H., Nguyen T.-T., Qu Y., Kim H.-M. 'Robustness improvement of superconducting magnetic energy storage system in microgrids using an energy shaping passivity-based control strategy'. *Energies*. 2017, vol. 10(5), p. 671.

[35]   Gil-Gonzalez W., Montoya O.D., Garces A., Espinosa-Perez G. 'IDA-passivity-based control for superconducting magnetic energy storage with PWM-CSC'. *Ninth Annual IEEE Green Technologies Conference (GreenTech)*; Denver, CO, 2017. pp. 89–95.

[36]   Montoya O.D., Gil-Gonzalez W., Serra F.M. 'PBC approach for SMES devices in electric distribution networks'. *IEEE Transactions on Circuits and Systems II*. 2018, vol. 65(12), pp. 2003–07.

[37]   Rodríguez A., Huerta F., Bueno E., Rodríguez F. 'Analysis and performance comparison of different power conditioning systems for SMES-based energy systems in wind turbines'. *Energies*. 2013, vol. 6(3), pp. 1527–53.

[38]   Muller S., Amrnann U., Rees S. 'New modulation strategy for a matrix converter with a very small mains filter'. *Power Electronics Specialists Conference (PESC)*; Acapulco, Mexico, 2003.

[39]   Hasanien H.M., Muyeen S.M. 'Design optimization of controller parameters used in variable speed wind energy conversion system by genetic algorithms'. *IEEE Transactions on Sustainable Energy*. 2012, vol. 3(2), pp. 200–08.

[40] Hasanien H.M., Muyeen S.M. 'A taguchi approach for optimum design of proportional-integral controllers in cascaded control scheme'. *IEEE Transactions on Power Systems*. 2013, vol. 28(2), pp. 1636–44.

[41] Hasanien H.M. 'Design optimization of PID controller in automatic voltage regulator system using taguchi combined genetic algorithm method'. *IEEE Systems Journal*. 2013, vol. 7(4), pp. 825–31.

[42] Hasanien H.M. 'A set-membership affine projection algorithm-based adaptive-controlled SMES units for wind farms output power smoothing'. *IEEE Transactions on Sustainable Energy*. 2014, vol. 5(4), pp. 1226–33.

[43] Wang S.C., Jin J.X. 'Design and analysis of a fuzzy logic controlled SMES system'. *IEEE Transactions on Applied Superconductivity*. 2014, vol. 24(5), pp. 1–5.

[44] Shi J., Tang Y., Yao T., Li J., Chen S. 'Study on control method of voltage source power conditioning system for SMES'. *IEEE/PES Transmission and Distribution Conference and Exhibition: Asia and Pacific*; Dalian, China, 2005.

[45] Sokolovsky V., Meerovich V., Spektor M., Levin G.A., Vajda I. 'Losses in superconductors under non-sinusoidal currents and magnetic fields'. *IEEE Transactions on Applied Superconductivity*. 2009, vol. 19(3), pp. 3344–47.

[46] Jing Shi, YuejinT., LiR., JingdongL., ShijieC. 'Discretization-based decoupled state-feedback control for current source power conditioning system of SMES'. *IEEE Transactions on Power Delivery*. 2008, vol. 23(4), pp. 2097–104.

[47] Nguyen T.-T., Yoo H.-J., Kim H.-M. 'Application of model predictive control to BESS for microgrid control'. *Energies*. 2015, vol. 8(8), pp. 8798–813.

[48] Ye Y., Kazerani M., Quintana V.H. 'Current-source converter based STATCOM: modeling and control'. *IEEE Transactions on Power Delivery*. 2005, vol. 20(2), pp. 795–800.

[49] Fuchs F.W., Kloenne A. *DC link and dynamic performance features of IPEMC*. IEEE; 2004.

[50] Monteiro V., Pinto J.G., Exposto B., Afonso J.L. 'Comprehensive comparison of a current-source and a voltage-source converter for three-phase ev fast battery chargers'. *2015 9th International Conference on Compatibility and Power Electronics (CPE)*; Costa da Caparica, Portugal, 2015. pp. 173–78.

[51] Deben Singh M., Mehta R.K., Singh A.K., Meng W. 'Integrated fuzzy-PI controlled current source converter based D-STATCOM'. *Cogent Engineering*. 2016, vol. 3(1), p. 1138921.

[52] Shi J., Tang Y., Ren L., Li J., Cheng S. 'Discretization-based decoupled state-feedback control for current source power conditioning system of SMES'. *IEEE Transactions on Power Delivery*. 2008, vol. 23(4), pp. 2097–104.

[53] Zhang C. *PWM converter and its control*. Beijing: Mechanic Industry Publishing House; 2003. p. 62.

[54]   Zhou S.P., Shi J., Jin T., He Q., Ren L., Tang Y.J. 'Improved state-feedback control for current source converter of SMES'. *Advanced Materials Research.* 2012, vol. 512–515, pp. 1049–54.

[55]   Singla H., Kumar A. 'Annual IEEE india conference (INDICON)'. Kochi, India, IEEE, 2012. pp. 92–286.

[56]   Jiayi L., Xinnan Z. 'SMES device and its power control strategy'. *IEEE Power Engineering and Automation Conference (PEAM)*; IEEE, 2011. pp. 244–47.

Chapter 4

Transient stability enhancement of power grids by superconducting magnetic energy storage

Mairaj ud Din Mufti[1], Hailiya Ahsan[1], and Abdul Waheed Kumar[1]

4.1   Introduction

A large chunk of researchers are still sceptical about the oddly high cost of energy storages. A Californian independent service operator, S. Berberich referred to energy storage systems (ESSs) as, 'It's good stuff, but it's expensive, and we've to find business cases'. However, with incoming efforts from certain power and energy monographs of the 2017 IEEE magazine issues, which are devoted to opening the door to energy storage challenges for future systems, the acceptance of such support systems is gaining importance. Despite the heated discussions rooted at a Switzerland seminar in the dawn of 2017, this ongoing decade has witnessed a colossal burgeoning of ESS implementation and experimentation. The naysayers could only cash on the economic side of such units as a disadvantage [1]. There is no denying the fact that ESSs that have been articulated to be the *holy grail* of modern power systems (*New York Times*, November 2010) are mostly economically not viable. It is, however, only as a matter of time and research that the still in *infancy* ESSs are not economically cheap. Had the scenario of renewable power penetration not assumed the power system stage, the call for ESS would still have been far away!

4.2   Transient stability issues in power grids

Power systems are being run closer to their stability limitations due to the expensive expense of transmission line additions. Transient stability has become a difficulty for sustaining the dependability of power systems as a result of this escalating strain on the system. The capacity of a power system to survive major disruptions such as

[1]National Institute of Technology Srinagar, Srinagar, Jammu and Kashmir, India

short circuit failures on transmission lines or generating unit outages is known as transient stability. Transient stability pertains to the ability of the systems' machines to sustain synchronism or a healthy profile when sudden intense disturbances, such as transmission cuts, loss of generation units/sources, abrupt load loss or line switching, perturb the system.

Since the 1920s, the stability of the power system has been recognized as a significant problem for secure system operation. The complexity of the power system is increasing worldwide with the permanent network extension and ongoing interconnections. The magnitude of this problem has been demonstrated by several global blackouts triggered by power system instability. The magnitude of this problem has been demonstrated by several global blackouts triggered by power system instability. Historically, transient instability has been the prevailing issue of stability in most systems and the focus of much of the industry's research on system stability. When power systems developed through continued growth in interconnections, the use of new technology and controls and increased activity under highly stressed conditions, numerous types of system instabilities have emerged. For example, voltage synchronization, synchronization of frequencies and inter-area oscillations have become more of a concern than in the past.

Transient stability study plays a vital role in power system stability investigations. It is common practice to operate power systems within the ambit of their steady-state operating points. Whenever a system is hit by an intense disturbance, it is prone to lose synchronism in the very first swing, if not provided with a proper transient support strategy. It is at this end, an energy storage like the SMES comes to rescue the troubled power system.

4.3 Transient stability improvement in conventional source-based power grids

SMES technology offers a remarkable potential to shape the transient behaviour of power systems in a beneficial way. SMES technology has the traits of a good energy storage device, primarily being good at efficiency, acting as a quick dynamic stability aid, equally good for all repetitive operations, intense response due to its unlimited number of charge and discharge cycles and so on. But in this era, where even the run-of-the-mill conventional power system poses to be a huge, complex nonlinear system, frequently characterized by bulk power transmissions over long distances and interconnected systems, SMES for power system control demands for a more advanced and intricate technology than other systems for momentary power system issues. For conventional power systems with no renewable penetration footprints, the transient stability studies are conducted in light of faults or other stringent contingencies only [2]. The element of intermittent power portions does not hold any share, whatsoever! So, rotor angle stability simply relates to the potential of synchronous machines in the power system to sustain synchronism when subjected to intense disturbances. It is concerned basically with upholding or maintaining the equilibrium between electrical and mechanical torque at each machine end in the system [3].

4.3.1   Elementary SMES modelling and control

As a promising storage technology, an SMES is a combination of an inductor (cooled at cryogenic temperatures), an extensive solid-state power conditioning devices unit (PCDU) and a supporting cryogenic refrigerator. A direct electric current flowing through the coil (I_{coil}) meets with no resistance, and therefore a zero virtual loss of power is seen. The coil stores energy as a magnetic field, not converting it to other forms. The PCDU consists of a pulsewidth-modulated voltage source converter (VSC) and a DC–DC chopper (two-quadrant), both connected by an insulated gate bipolar transistor. The former acts as an interface (buffer) between the bus and the coil, while the latter classifies for an *intrinsic control action* determining the voltage to be applied across the coil (V_{coil}) required for relevant charging/discharging [4]. The SMES configuration taken up in this chapter is illustrated in Figure 4.1(a), depicting a Y-Δ transformer, followed by a VSC and a DC–DC chopper circuitry, connected via a DC link. This combination is completed by the superconducting inductor or coil. The VSC plays its role as the grid-side converter, regulating the link voltage around its nominal value. Figure 4.1(b) gives the SMES modelling and control strategy, based on a *P–f* regulation method. The SMES is presented as an active current source having the capability of exchanging active power with the system. The active power modulation of SMES can be related in Figure 4.1(a), where S_1 and S_2 represent the IGBT switches and D_1 and D_2 represent the diodes. Three regions of operation are possible for the modus operandi of the chopper depending on the value of duty cycle, D [4]. The timing diagrams pertaining to these regions of operation are shown in Figure 4.2.

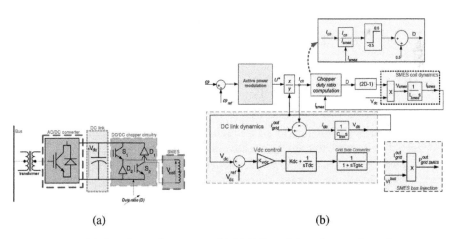

(a)   (b)

Figure 4.1   *(a) SMES configuration; (b) SMES model and control: real power modulation P–f loop*

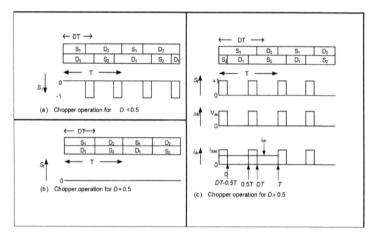

Figure 4.2   *Details about the ON positions of various switching devices of the*
chopper for varying duty cycle values

The role of the two-quadrant chopper is significant because it generates the duty
ratio (*D*) for the switching operation, entirely from the reference power command.
Detailed duty ratio timing diagrams can be located in Reference 4. The resulting
chopper current (I_{chop}) is deduced from the corresponding coil current (I_{coil}) and duty
ratio (*D*) as

$$
I_{chop} = \begin{cases} I_{coil}, & 0 \le t \le DT - T/2 \\ 0, & DT < t \le T/2 \\ I_{coil}, & T/2 < t \le DT \\ 0, & DT < t \le T \end{cases}
\tag{4.1}
$$

As incorporated in Figure 4.1(b), the duty ratio and average chopper current are
related as

$$
I_{chop} = \frac{1}{T}\left[\int_0^{DT-\frac{T}{2}} I_{coil}dt + \int_{DT}^T I_{coil}dt \right]
\tag{4.2}
$$

Thus, the frequency regulatory loop [see Figure 4.1(b)] merges within an underlying
chopper action control, to track an effective SMES active power trajectory.
    As a magnetic ESS, an SMES absorbs/delivers real power to (or from) the net-
work, depending on the voltage applied across the magnetic coil [5]. It is a DC-
driven device, in which the duty ratio determines the nature of the applied voltage
(positive/negative) across the coil. By virtue of a substantial power conditioning support, which comprises a VSC and a two-quadrant DC–DC chopper, it also ensures
a bidirectional reactive and active power control, respectively [6]. Typically, as a
source/sink of huge power bursts, an SMES is characterized by a significantly low

discharge time (less than 80 s). A typical model of an SMES system can be seen in Figure 4.1(a). It comprises an inner loop dynamics, which encompasses a duty ratio determining chopper action and the magnetic coil voltage scenario. Mathematically,

$$\left\{ 2 \underbrace{\left(0.5 + \frac{I_{chop}}{I_{coil}} \right) - 1}_{\text{Duty ratio (D)}} \right\} V_{dclink} = L_{smes} \frac{dI_{coil}}{dt} \tag{4.3}$$

To maintain an effectively longer time span of the power conditioning support, the DC link voltage is also monitored to yield a smooth link voltage profile:

$$I_{chop} + C_{link} \frac{dV_{dclink}}{dt} = I_{grid} \tag{4.4}$$

4.4 Transient stability improvement in renewable energy source-based power grids

Any power system anticipating intermittent power outputs of wind energy sources unavoidably faces excursions in power system swings. These exogenous perturbations result in the degradation of the stability of the gross system [7]. A superconducting magnetic energy storage has repeatedly come to the aid of such system abnormalities and proved its mettle in mitigating/alleviating the same [8,9]. The SMES applied science came to the limelight in the 1970s to mitigate the oil crisis, arising due to the impractical fuel prices. An SMES has been reckoned as an efficient damping device, capable of performing a swift dual power exchange, i.e. a simultaneous real and reactive power supply or storage [10, 11]. The routine requirement of meeting increased load demand in common power systems is mostly met by enhancing the overall installed capacity by availing wind energy sources.

Potency of the SMES unit to handle the uncertainties is determined by the control strategy that has been used to handle both real and reactive power flow in a power system comprising of intermittent power sources [12]. The performance of fixed gain controllers worsens due to nonlinearities and uncertainties occurring in the system. Moreover, they are not capable of handling the constraints that are an integral part of any realistic system. Power system being a nonlinear time-varying, multiple-input–output system has many closed loops. So, for modelling a particular system, idealized assumptions and simplifications are considered, and as a result, the models that are obtained do not adequately capture the physical system. From a control design point of view, a control system is to be designed to maintain satisfactory system performance under various uncertainties. Fixed gain controllers require accurate system modelling, i.e. system is based on a priori information and may fail while handling an exogenous disturbance of a higher magnitude. For a fixed gain controller, accurate system modelling is required to reduce the system performance error. So, there exists a direct tradeoff vis-a-vis performance and uncertainty. Therefore, for a high level of uncertainty, fixed gain controllers fail to maintain acceptable system performance. They do not consider the change in dynamics and structure. On the other hand, an intelligent controller does not depend on the system modelling excessively and improves itself for any dynamic or adverse condition. Such controllers have a fixed structure with a self-tuning feature that adjusts the system parameters automatically while dealing with the parametric uncertainty.

The International Federation on Automatic Control (Industry's Committee) conducted a poll of its members in 2015 to determine the influence of advanced control techniques on industrial applications. If a given controller had resulted in significant benefits to one or more industry sectors and was successfully deemed a standard practise, the impact was assumed to be considerable. The proportional integral derivative control, also known as the PID scheme, nailed the task. In the literature, additional techniques that are more complex and resilient than PID control have been offered. These techniques may be deterministic or model dependent, or even heuristically developed. Techniques such as the model predictive control, sliding mode control, fuzzy logic control and H infinity control have proved exemplary robustness in comparison to the PID domains. Through this chapter, we intend to portray the elementary modelling and control of SMES for transient stability enhancement in modern-day power systems. The goal is for the constructed models to perform admirably, make reasonably accurate forecasts and lead to competent engineering device design. The goal of this chapter is to classify the dynamic behaviour of various SMES control strategies under various contingencies and system scenarios using a thorough stochastic approach. While we are well aware that the industry is always eager to implement a new control approach as long as it is a PID, these techniques, while more involved, offer certain advantages that are not always possible to obtain with traditional PID control.

4.4.1 SMES control algorithms for transient stability enhancement

4.4.1.1 SMES with derivative control

The SMES control strategy will focus on a concerted active-reactive power regulation mechanism. The SMES units' real power regulation is based on a frequency control approach, whilst the reactive power modulation is based on voltage preservation (Q–V control) [13]. In the P–f control, two different metrics have been endorsed: generator frequency deviation (*Delta* control) and derivative control (determined through the rate of change of frequency at the generator). These changes are caused by an immediate misalignment across sources and loads as a result of recurrent power failures [14].

Contingent on the converter size, the baseline demands, P^* and Q^*, may not be practical to meet, especially if the system's volt–ampere demand exceeds that of the SMES. As a result, adding restrictions to the reference commands yields a better power requirement approximation. Mathematically,

$$P_{dem} = \left[\frac{P^*}{\sqrt{P^{*2} + Q^{*2}}}\right] V_{bus} I_{dc} \qquad Q_{dem} = \left[\frac{Q^*}{\sqrt{P^{*2} + Q^{*2}}}\right] V_{bus} I_{dc} \qquad (4.5)$$

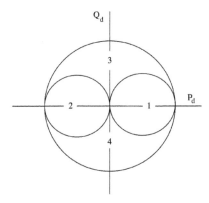

Figure 4.3   SMES P–Q four-zone control domain

Here, P_{dem} and Q_{dem} are the realizable power demands of the system, when the steady-state operating point gets perturbed. V_{bus} and I_{dc} are the SMES bus voltage and coil current in p.u. P_{dem} and Q_{dem} are actually the active and reactive power demands of the system, which are used to calculate the firing angles of SMES converters. Two sets of gate turn-off thyristors (GTOs) are used, where Δ–Y and $\Delta\Delta$– transformers make a 12-pulse arrangement. The DC ripple voltage is reduced because the high-side voltage and current in $Y\Delta$–/Δ–Y transformers lead the low-side voltage by 30° for positive sequence, and the high-side voltage and current lag the low side by 30° for negative sequence. Since two sets of GTO converters are used, the possible regions of firing angles of the converter are from 0° to 360°. At first, the region where the (P_{dem}, Q_{dem}) point lies is located using Figure 4.3. Next, the calculation of firing angles $alphas_1$ and $alphas_2$ is done using (4.6) to (4.9) along with Table 4.1. In case $P_{dem} = Q_{dem} = 0$, an infinite number of combinations of $alphas_1$ and $alphas_2$ exist and one of the solutions is $alphas_1 = 90°$ and $alphas_2 = 270°$deg.

This quick firing angle control helps the SMES power output to track the power demand continuously. The four-zone P–Q spread, for firing angle control is depicted using the control domains of GTO converters as in Figure 4.3.

$$\alpha_1 = \cos^{-1}\left[\frac{1}{2E_{dc0}I_{dc}}[P_{dem}\pm Q_{dem}Q_d]\right] \tag{4.6}$$

Table 4.1   Power demand-based firing angle calculation

Zone	α_1	α_2	Complex sign
1	(4.6)	(4.9)	Lower side
2	(4.6)	(4.8)	Upper side
3	(4.6)	(4.9)	Upper side
4	(4.7)	(4.9)	Upper side

$$\alpha_1 = 2\pi - \cos^{-1}\left[\frac{1}{2E_{dc0}I_{dc}}\left[P_{dem} \pm Q_{dem}Q_m\right]\right] \tag{4.7}$$

$$\alpha_2 = \cos^{-1}\left[\frac{1}{2E_{dc0}I_{dc}}\left[P_{dem} \mp Q_{dem}Q_m\right]\right] \tag{4.8}$$

$$\alpha_2 = 2\pi - \cos^{-1}\left[\frac{1}{2E_{dc0}I_{dc}}\left[P_{dem} \mp Q_{dem}Q_m\right]\right] \tag{4.9}$$

$$\text{where} \quad Q_m = \left[\sqrt{\frac{4E_{dc0}{}^2 I_{dc}{}^2 - (P_{dem}{}^2 + Q_{dem}{}^2)}{P_{dem}^2 + Q_{dem}^2}}\right]$$

The SMES coil's energy level feedback is used to efficiently realize a realistically rational range of device performance. This SMES state of charge level is constantly monitored in order to avoid energy saturation. Constraints on the energy transfer between the SMES unit and the grid are applied to maintain the energy exchange between the SMES unit and the grid within the prescribed limits set by the power conversion system. For continuous control, a limit should be imposed on the minimum stored energy level of SMES. A restriction on the minimum stored energy level of SMES is enforced for continuous control. For profitable operation and constant management, the SMES current [15] /energy [14] must always stay well within the upper/lower ranges. The SMES coil energy is approximated as

$$E_{sm} = \int P_{smes}\, dt \tag{4.10}$$

This energy feedback component is mixed with the frequency regulated resultant, to generate the optimal power reference command, P^*. Mathematically, the energy feedback regulation is incorporated as

$$E_k = k_e\left[\int_{t_0}^{t_{sim}} P_{smes}\, dt - E_{int}\right] \tag{4.11}$$

Here, E_{int} is the initial charge level of the SMES coil. The energy level is monitored continuously to keep the energy within the vicinity of the nominal energy value.

A mixed integral squared error (ISE) is carved out of the deviations in frequencies ($\Delta\omega$) and voltages of the SMES clad bus; along with SMES SoC deviations ($\Delta\,SoC_{sm}$), which serves as the cost function to estimate the optimal inertial gain values. P–f and Q–V coupling is adhered to, in cost function creation. Optimal values of K_{pf}, K_{qv} and K_e are obtained by minimizing the resulting cost function, subject to certain constraint bounds [16, 17].

Mathematically, the objective function is created as

$$C_f(K_{pf}, K_{qv}, k_e) = \int_{t_0}^{t_{sim}} \left[(A\Delta\omega^2 + B\Delta V^2) + C\Delta SoC_{sm}^2\right] \tag{4.12}$$

$$\text{subject to}: K_{pf}^{min} \leq K_{pf} \leq K_{pf}^{max}, k_{qv}^{min} \leq K_{qv} \leq K_{qv}^{max}, k_e^{min} \leq k_e \leq k_e^{max} \tag{4.13}$$

Here, t_{sim}–t_0 is the simulation span, and A, B and C are the scaling weights, deciding the damping share for the modulation strategy, because P–f control delivers dominantly in damping [18]. Equation (4.12) deliberates towards maintaining the time-domain dynamic stability of the system.

4.4.1.2 SMES as a virtual synchronous generator

In conventional power plants, the alternators provide inertia to the grid through their rotating parts. This inertia helps to regulate grid frequency by avoiding frequency deviation from the regular frequency operation. But, these power plants are responsible for greenhouse gas emissions. To reduce greenhouse emissions, the integration of renewable energy systems (RESs) is encouraged. However, these RESs have either small or no rotating part. Therefore, they reduce the inertia of the grid dramatically leading to frequency/voltage instability as compared to the system using synchronous generators [19, 20]. The critical clearing time reduces because of reduction in effective network inertia due to asynchronous generation [21]. The most significant challenge for the integration of RESs is frequency regulation. To imitate the behaviour of a traditional alternator, a virtual synchronous generator (VSG) is used. The concept of VSG aims to emulate the behaviour of synchronous generator in order to stabilize the power system [22]. A typical VSG is a combination of an energy storage, an inverter and any control mechanism. The energy storage may be an ultrabattery, a supercapacitor, a flywheel, an SMES, or a battery. Being a high power density storage device, the SMES technology is deemed fitter for short yet rapid power bursts for load levelling in power systems with critical loads and can act as a spinning reserve whilst contingencies, as its discharge time is typically lesser than 100 s. The reference output power command of SMES-VSG is given by

$$P_{VSG} = k_I \frac{d\Delta\omega}{dt} + k_D \Delta\omega + k_C(I_{sm}^0 - I_{sm})$$ (4.14)

where, $\Delta\omega = \omega - \omega_0$, ω_0 is the nominal frequency of the system.

The first term in (4.14) emulates the power that is absorbed/released by positive or negative frequency deviation rate $(\frac{d\Delta\omega}{dt})$[23]. K_I is the inertia emulating coefficient and determines response to high ROCOF values. Power will be exchanged only during transients because the change in frequency is zero in steady state. After a disturbance, if the frequency of the system decreases, SMES-VSG starts to supply power till $(\frac{d\Delta\omega}{dt}) = 0$.

If, on the other hand, the frequency of the system increases, SMES-VSG starts to absorb power till $(\frac{d\Delta\omega}{dt}) = 0$. The second term emulates the damper winding effect of SG and helps to restrain maximum frequency deviation.

The value of K_C is properly chosen. The large value of K_C will make SMES-VSG unit ineffective in reducing the frequency deviations. On the other hand, a very small value of K_C will not allow the SMES coil to return to its nominal value [24].

The reference input to the SMES is calculated by

$$P_{SMES}^* = \Delta\omega + k_C(I_{SMES}^0 - I_{SMES})$$ (4.15)

The reference input to the SMES when it is operating as VSG is calculated by

$$P_{VSG} = k_I \frac{d\Delta\omega}{dt} + k_D \Delta\omega + k_C(I_{SMES}^0 - I_{SMES})$$ (4.16)

where I^0_{SMES} is the nominal current of the SMES unit; $\Delta\omega$ is the frequency deviation and is calculated using the concept of centre of inertia [25]. For NETS and NYPS, ω_{COI} is calculated as

$$\omega_{COINETS} = \frac{\sum_{i=1}^{9} \omega_i H_i}{\sum_{i=1}^{9} H_i}, \qquad \omega_{COINYPS} = \frac{\sum_{i=9}^{13} \omega_i H_i}{\sum_{i=9}^{13} H_i}$$

where ω_i is the ith generator speed and H_i is the inertia constant of the ith generator.

4.4.1.3 Discrete predictive control of an SMES

The set points of SMES control are issued by a discrete predictive controller (DPC). This allows the participation of SMES by an immediate exchange of real and reactive power while taking into control the hardware/operational constraints as imposed by the power electronic circuitry (the grid-side converter, machine-side converter and the intermedial DC link). The DPC is applied to a state-space model arrangement of the SMES energy storage. The SMES reference power command (P^*_{SMES}) is constrained by a dynamic constraint window (DCW), and step-ahead estimates are made about the SMES energy (E_{SMES}). SMES inner loop dynamics is approximated by a lumped second-order system (time constants, T_1 and T_2), restrained by temporary wind-up limiters [26]. To sustain a continuous operation, the SMES state of charge is checked and a corresponding term is amalgamated in the active power regulation loop of the SMES.

A pulse transfer function is utilized to approximate the SMES energy level (E_{SMES}) in its discrete form as the area under the curve of reference SMES power command (P^*_{SMES}) [27]. Mathematically, we cast it as

$$W(z) = \frac{E_{SMES}(z)}{P^*_{SMES}(z)} = \mathscr{L}\left[J_{zoh}(s) \frac{1}{s(1+sT_1)(1+sT_2)} \right] \tag{4.17}$$

where J_{zoh} accounts for the sample and hold of data, with sample time (T) being 0.1 s.

It is evident from (4.17), that the predictive control model can be split into two distinct domains, a continuous domain and a discrete domain. The state equation sets for the continuous portion of (4.17) can be defined as [27]

$$\begin{cases} \dot{X}(t) = AX(t) + BU(t) \\ Y(t) = CX(t) \end{cases} \tag{4.18}$$

A DCW (within $\pm P_{dcw}$) is applied to the SMES power command, over the prediction horizon (N_{end}), to prevent its excess charge/discharge and facilitate optimal sizing of the SMES. The control vector P^*_{SMES} is kept under control over the entire band of control horizon (N_{con}) [see (4.19)] [24]:

$$
\begin{bmatrix}
1 & 0 & 0 & 0 & \cdots & \cdots & 0 \\
0 & 1 & 0 & 0 & \cdots & \cdots & 0 \\
\vdots & \vdots & \vdots & & & & \vdots \\
0 & 0 & 0 & \cdots & 1 & 0 & \cdots \\
\vdots & \vdots & \vdots & & & & \vdots \\
0 & 0 & 0 & \cdots & \cdots & 0 & 1 \\
-1 & 0 & 0 & 0 & \cdots & \cdots & 0 \\
0 & -1 & 0 & 0 & \cdots & \cdots & 0 \\
\vdots & \vdots & \vdots & & & & \vdots \\
0 & 0 & 0 & \cdots & -1 & 0 & \cdots \\
\vdots & \vdots & \vdots & & & & \vdots \\
0 & 0 & 0 & \cdots & \cdots & 0 & -1
\end{bmatrix}
\begin{bmatrix}
P^*_{smes}(k) \\
P^*_{smes}(k+1) \\
P^*_{smes}(k+2) \\
\vdots \\
P^*_{smes}(k+N_{con}) \\
\vdots \\
P^*_{smes}(k+N_{end})
\end{bmatrix}
\leq
\begin{bmatrix}
P_{dcw} \\
P_{dcw} \\
P_{dcw} \\
\vdots \\
-P_{dcw} \\
-P_{dcw} \\
\vdots \\
-P_{dcw}
\end{bmatrix}
\tag{4.19}
$$

Mathematically, a discrete-time state-space model is developed from continuous-time state equations in (4.18). In order to predict the flywheel energy for future states, the resulting third-order system can be solved for the state transition matrix, $\Phi(t)$ [27]. The derived discretized model of the SMES system for DPC application is given as

$$
\begin{cases}
X(k+1) = A_z X(k) + B_z U(k) \\
Y(k) = C_z X(k)
\end{cases}
\tag{4.20}
$$

$$
\text{where }
\begin{cases}
A_z = \left[\int_0^T \Phi(\tau)d\tau \right]_{T=0.1s} \\
\\
B_z = \left[\int_0^T \Phi(\tau)d\tau \right]_{T=0.1s} * B
\end{cases}
\tag{4.21}
$$

The aforementioned continuous-time (4.18) to discrete-time system formulations (4.20) result in the following DPC-oriented state matrix, A_z, as given in (4.22) below:

$$
P^*
\begin{bmatrix}
e^{-mT} - e^{-nT} & T_1 e^{-mT} - T_2 e^{-nT} + g & 0 \\
\\
T_2 e^{-nT} - T_1 e^{-mT} - g & R & 0 \\
\\
Q & T_2(e^{-nT} - 1) + T_1(1 - e^{-mT}) + (T_1 + T_2)Q & T/P
\end{bmatrix}
\tag{4.22}
$$

$$\begin{cases} m = 1/T_1, \quad n = 1/T_2, \quad g = (T_2 - T_1), \\ \\ P = \frac{1}{T_1 - T_2} \quad \text{and} \quad Q = T_1^2(e^{-mT} - 1) + T_2^2(1 - e^{-nT}) + T \\ \\ R = e^{-mT} - e^{-nT} + (T_1^2 - T_2^2) + (T_1 + T_2)(T_2 e^{-nT} - T_1 e^{-mT}) \end{cases}$$

By virtue of the continuous-time state equations evolving from (4.17) to (4.18), DPC models input matrix is a function of the state matrix and is put forth as

$$B_z = P \begin{bmatrix} e^{-mT} - e^{-nT} \\ \\ T_2 e^{-nT} - T_1 e^{-mT} + (T_1 - T_2) \\ \\ T_1^2(e^{-mT} - 1) + T_2^2(1 - e^{-nT}) + T \end{bmatrix} \tag{4.23}$$

The system output, E_{smes}, is predicted as

$$E_{smes}(k+1) = \begin{bmatrix} 0 & 0 & 1 \end{bmatrix}^T X(k) \tag{4.24}$$

Equation (4.25) gives the discrete predictively controlled SMES model in numerical state-space notations:

$$A_z = \begin{bmatrix} 0.0837 & -0.9822 & 0 \\ -0.0098 & 0.8751 & 0 \\ 0.1566 & 1.0872 & 0.1 \end{bmatrix}, \quad B_z = \begin{bmatrix} 0.8437 \\ -0.0098 \\ 0.1566 \end{bmatrix} \tag{4.25}$$

4.4.1.4   Constrained neural-adaptive predictive control of SMES

A regulating variable $y(k)$ is meticulously chosen, so that the reference power command P^*_{smes} conforms to the control objectives (like regulation of frequency, SMES state of charge, etc.). In general, the power system with SMES unit is a multi-input, multi-output nonlinear system, which can be expressed as

$$y(k+1) = f(y(k), y(k-1), ..., y(k-n), u(k), u(k-1), ..., u(k-m)) \tag{4.26}$$

where $y(k)$ and $u(k)$ are the scalar output and the input of the system, respectively, f is the nonlinear function to be approximated by the neural network and n and m are the known structure orders of the system. A two-layer neural network is utilized to contemplate the dynamics between $y(k)$ and the power command, P'_{smes}, the activation functions being hyperbolic tangent for the first layer and linear for the second layer. The neural model for the system (4.26) can be imparted as [28]

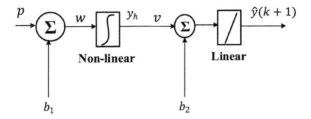

Figure 4.4   Neural network structure

$$\hat{y}(k+1) = \hat{f}(y(k), y(k-1), ..., y(k-n), u(k), u(k-1), ..., u(k-m)) \qquad (4.27)$$

where $\hat{y}(k+1)$ is the one-step-ahead prediction of $y(k)$ and $\hat{f}$ is the approximation of f. Using the neural network model illustrated in Figure 4.4, (4.27) can be written as

$$\hat{y}(k+1) = v\left[tanh(wp + b_1)\right] + b_2 \qquad (4.28)$$

where $p = [y(k), y(k-1), ..., y(k-n), u(k), u(k-1), ..., u(k-m)]$ is the input to the neural network and v, w, b_1, b_2 are the weights and biases matrices, which are updated by means of the back-propagation algorithm, considering the disturbances in order that the performance index J_1 given by (4.29) is minimized:

$$J_1 = \frac{1}{2}[\hat{y}(k) - y(k)]^2 \qquad (4.29)$$

Subsequently, to generate the power command, P'_{smes} (see Figure 4.6), the performance index J_2 given in (4.30) is minimized:

$$J_2 = \frac{1}{2}\hat{y}^2(k+1) + \frac{1}{2}\lambda u^2(k) \qquad (4.30)$$

where λ is introduced as a penalty factor on control, and in case $\lambda = 0$, the control scheme eventuates in inverse control, making its application incongruous for the systems exhibiting an unstable or poorly damped inverse. However, the discretization of continuous time models often results in such a situation despite of the system being stable. $u(k)$ is recursively computed by using an elementary gradient descent rule in order to minimize J_2:

$$u(k+1) = u(k) - \eta\frac{\partial J_2}{\partial u(k)} \qquad (4.31)$$

where $\eta > 0$ is the learning rate. The controller performance relies on the estimation made by the neural network. Therefore, it is imperative that $\hat{y}(k+1)$ approximates the actual output $y(k+1)$ asymptotically. This is achieved by training the neural network online. Differentiating J_2 with respect to $u(k)$

$$\frac{\partial J_2}{\partial u(k)} = \hat{y}(k+1)\frac{\partial \hat{y}(k+1)}{\partial u(k)} + \lambda u(k) \qquad (4.32)$$

where $\frac{\partial \hat{y}(k+1)}{\partial u(k)}$ is termed as neural network gradient and can be determined by employing the network structure given by (4.28), as jotted in (4.33) and (4.34):

$$\frac{\partial \hat{y}(k+1)}{\partial u(k)} = v\left[sech^2(wp + b_1) \right] w \frac{dp}{du} \tag{4.33}$$

$$\frac{dp}{du} = [0, 0, ..., 0, 1, 0, ..., 0]' \tag{4.34}$$

Therefore, the control signal can be interpreted as

$$u(k+1) = (1 - \eta\lambda)u(k) - \eta v\hat{y}(k+1)\left[sech^2(wp + b_1) \right] w \frac{dp}{du} \tag{4.35}$$

For real-time control, the control and identification processes explained can be realized by using *s-function* in MATLAB®/Simulink®. The compendium of *s-function* operation is as follows.

1. Evaluate $\hat{y}(k)$.
2. Measure $y(k)$.
3. Update weights and biases employing back-propagation algorithm to minimize J_1.
4. Compute $\hat{y}(k+1)$ using (4.28).
5. Compute new control signal $u(k+1)$ using (4.35) and feed to the SMES PCS.

Generation of the energy level constraints

The output power of SMES is the output of a first-order lag compensator that represents the SMES control converter:

$$P_{smes} = \frac{1}{1 + sT_E} P^*_{smes} \tag{4.36}$$

Using this fact, an integrator is cascaded with the first-order lag block and a control-oriented discrete-time model of SMES is developed to procure its energy level constraints.

The resulting second-order plant can be represented in continuous-time state variable form given in (4.37):

$$\dot{X} = A_c X + B_c U \tag{4.37}$$

where

$$A_c = \begin{bmatrix} 0 & 1 \\ 0 & \frac{-1}{T_E} \end{bmatrix} ; B_c = \begin{bmatrix} 0 \\ \frac{1}{T_E} \end{bmatrix}$$

The discrete-time model of the system in (4.37) can be represented as

$$\begin{bmatrix} x_1(k+1) \\ x_2(k+1) \end{bmatrix} = \begin{bmatrix} a_{11} & a_{12} \\ a_{21} & a_{22} \end{bmatrix} \begin{bmatrix} x_1(k) \\ x_2(k) \end{bmatrix} + \begin{bmatrix} n_1 \\ n_2 \end{bmatrix} u(k) \tag{4.38}$$

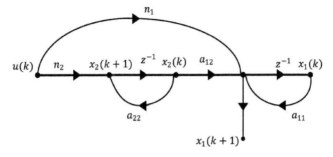

Figure 4.5   Signal flow graph for energy level prediction

where $x_1(k) = E_{smes}(k)$, SMES energy level at the kth instant $x_2(k) = P_{smes}(k)$, actual power delivered to SMES. The system in (4.38) is cast into a signal flow graph as portrayed in Figure 4.5, to acquire the one-step-ahead prediction of energy level in terms of SMES power command P^*_{smes}.

Applying Mason's gain formula while choosing $x_1(k+1)$ and $u(k)$ as output and input nodes, respectively, we get

$$\frac{x_1(k+1)}{u(k)} = \frac{n_1(1 - a_{22}z^{-1}) + n_2a_{12}z^{-1}}{1 - (a_{11} + a_{22})z^{-1} + a_{11}a_{22}z^{-2}} \tag{4.39}$$

which yields

$$x_1(k+1) = n_1u(k) + (n_2a_{12} - n_1a_{22})u(k-1) + (a_{11} + a_{22})x_1(k)$$
$$-a_{11}a_{22}x_1(k-1) \tag{4.40}$$

Articulating (4.40) in terms of physical variables, we get

$$E_{smes}(k+1) = n_1P^*_{smes}(k) + (n_2a_{12} - n_1a_{22})P^*_{smes}(k-1) + (a_{11} + a_{22})E_{smes}(k)$$
$$-a_{11}a_{22}E_{smes}(k-1) \tag{4.41}$$

Equation (4.41) gives the desired prediction-based discrete control model of SMES, and the dynamic power thresholds are interpreted as

$$P^*_{smes}(k)_{min} = C_1E_{min} - C_2P^*_{smes}(k-1) - C_3E_{smes}(k) + C_4E_{smes}(k-1) \tag{4.42}$$

$$P^*_{smes}(k)_{max} = C_1E_{max} - C_2P^*_{smes}(k-1) - C_3E_{smes}(k) + C_4E_{smes}(k-1) \tag{4.43}$$

where $\begin{cases} C_1 = \dfrac{1}{n_1}, C_2 = \dfrac{n_2a_{12} - n_1a_{22}}{n_1}, C_3 = \dfrac{a_{11} + a_{22}}{n_1}, C_4 = \dfrac{a_{11}a_{22}}{n_1} \\ a_{11} = 1, a_{12} = T_E(1 - e^{\frac{-T}{T_E}}), a_{21} = 0, a_{22} = e^{\frac{-T}{T_E}}, n_1 = T - T_E + T_Ee^{\frac{-T}{T_E}} \\ n_2 = 1 - e^{\frac{-T}{T_E}} \end{cases}$

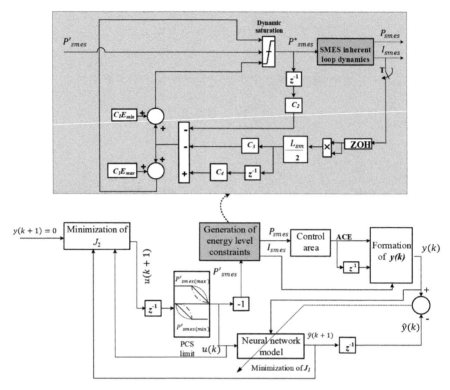

Figure 4.6   Constrained neural-adaptive predictive control architecture

E_{min} and E_{max} represent the minimum and maximum energy levels of SMES and are calculated as

$$E_{max}^2 - E_0^2 = E_0^2 - E_{min}^2 \qquad (4.44)$$

where $E_0 = \frac{1}{2}L_{sm}I_{smes}^{o}{}^2$ is the nominal energy level of SMES. The architecture of constrained neural-adaptive predictive control is delineated in Figure 4.6.

4.4.1.5   Heuristic fuzzy logic control

Fuzzy logic has often been termed as far from complete, due to the in-existence of a systematic procedure. The sheer absence of a systematic outlined procedure for fuzzy logic primarily hampers its application to any time-varying nonlinear system. The designers' iterative approach of performing trial and error is not only time consuming but it also shows an element of sensitivity to variation in system operating points, just like other conventional PI-driven control systems. Overall, the missing link remains the ambiguity pertaining to the creation of a complete rule base and absence of a unique set of overlap and membership functions [29–31]. The attempt is to relax the constraints of the conventional fuzzy logic application, by allowing an intelligent control method by using an evolutionary algorithm to optimally dispatch the fuzzy template [32]. Genetic algorithms (GAs) are particularly suited to problems with variables, either integer valued or continuous, being constrained within some upper and lower thresholds. Also, the ability of GAs to surf the search paradigm in parallels, without any requirement of the optimization function to be differentiable or smooth, makes it the most amenable option for our problem [33]. A custom GA function is coordinated with mixed integer programming, employing intcon as the integer constrained vector prompt.

The standard fuzzy template is obtained using mixed integer programming-based mathematics, which is optimally tuned using GAs. The target lies in establishing the component values for the control strategy to yield a structured approach for the fuzzy controller. Practically, infinite possible outcomes of chromosome strings carrying encoded fuzzy information are possible, which may vary with the variations in the system operating state. The target is to find the best values of the bits in the chromosome array/string to accord all-inclusive favourable results matching the steady-state values. In an optimization sense, it is attempted to minimize the height of the area under the curve of an integral squared quantity. As it is, an integer-constrained optimization problem comes at hand, because the fuzzy rule base features crisp quantified levels, emanating from a pre-defined universe of discourse. However, the other share of the string is clearly not integer constrained due to the flexibility granted to membership function spreads as well as the series aligned fuzzy weights. Thence, a mixed integer constrained chromosome string is to be tackled to yield an optimized fuzzy template [34].

A few points worth noting here are as follows:

1. The absence of nonlinear and linear equality constraints in the problem facilitates mixed-integer optimization, as the equality and integer constraints cannot co-exist for mixed integer programming.
2. A dichotomy exists between the fitness and penalty function concerning the non-feasibility of the search variables based on integer constraints, in which case the penalty function retains the value of the highest fitness function among the search variables of the population. The dichotomy seizes if and only if the search variable is feasible.
3. Tightly binding the upper and lower thresholds, for all the search variables so as to reduce the heuristic search space and quick convergence of GA.
4. The integer constraints of the mixed integer optimization are met by GA built creation, crossover and mutation.

An encoded chromosome string is designed with the chromosome apices representing the input as well as output membership spread, symmetric fuzzy logic rules and series fuzzy weights. A compact chromosome string template is designed using six membership value bits (four inputs, two outputs) and 13 mirror image logic rules. In this study, symmetric trapezoidal membership functions are chosen as fuzzy system inputs, while symmetric singletons are designated as output membership spreads. This is primarily done to restrict the chromosome string to a smaller order of 19. A triangular cum trapezoidal symmetric membership function is characterized by five distinct nodes as [(−*b*, −*a*, *0*, *a*, *b*); (−*d*, −*c*, *0*, *c*, *d*)] for the two inputs. Similarly, the output singletons are classified along (−*f*, −*e*, *0*, *e*, *f*) as crisp locus points. In this work, a medium-sized universe of discourse for all input/output selections is adhered to and classified under five labels, viz. negative big (NB), negative (N), zero (Z), positive (P) and positive big (PB). The linguistic labels NB, N, Z, P and PB are operated as 1, 2, 3, 4 and 5, respectively, in numerical denominations. These 5 × 5 rules are split symmetrically along the mid-locus as mirror images, thereby reducing the length of the encoded string to 13 only.

First, the designing summons certain bounds on each of the template variables. Our variables will be the indices for the fuzzy rule base vector, hence fixing the lower bound always to 1 (i.e. the lowest possible index for the vectors). The upper bounds at the same time shall be the length of the available bit options. Because we are interested in a symmetric rule base, a mirror image segregation shall create a space for 13 possible rule loci, out of a spectrum of 25 distinct options. Clearly, each of the first 13 entries of the vector shall have the upper bounds set to 5 and are constrained to have integer values, due to the available options from the universe of discourse. The other sections of the vector, however, enjoy all real values, defined within realistic upper and lower bounds.

Remark 4.1: The problem is set up and defined in MATLAB®.

4.4.1.6　Parametric fuzzy predictive control

Parametric fuzzy predictive control as the name suggests is a selective combination of predictive and fuzzy approaches. The parameters are explicitly obtained with reduced computational effort and less memory requirement. A discrete predictive control scheme is designed and applied for real power modulation of SMES and the aforementioned heuristic fuzzy logic control caters to volt-var control (or reactive power modulation). Figure 4.7 gives the control architecture of an SMES being run using PFPC.

4.5　Case studies

4.5.1　Case A

This section explores the impact of a derivative control-driven SMES on transient stability. Figure 4.17 shows the modified version of the benchmark Western System Coordinated Council 9-bus system. The studied system consists of three synchronous generators and an additional 50-MVA wind farm at bus 4, to consider the effect of wind penetration. Each generator is modelled using the *two-axis model* terminology, where all the flux linkage and damper winding dynamics are assumed considerably fast. Each alternator is equipped with an efficient automatic generation control and IEEE type-I excitation as the automatic voltage regulator [35].

The power swing variables of frequency, voltage and power are closely followed to assess the dynamic performance of the system, with and without SMES units in place. The transient behaviour of the system is also analysed by tracking

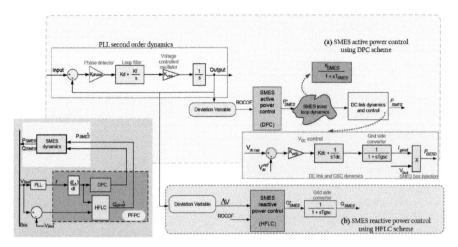

Figure 4.7   *(a) PFPC control architecture depicting reactive and active power control of SMES. (b) SMES compact model with inner loop dynamics appoximated by a first-order delay and PI control-based DC link voltage regulation*

damping ability in machine angle oscillations, when a disturbance hits the system. Two different kinds of contingencies of varying magnitudes (*a 250 ms, 3-ϕ fault on the line connecting buses 5–7 at t = 1 s, and outage of the line connecting buses 8–9 at t = 2 s*) are applied to the system, so as to investigate the effectiveness of the presented SMES control schemes.

Evidently, the derivative and Δ control techniques reduce the ISE by over 79 and 72%, respectively, when compared with the system devoid of any SMES units. Table 4.2 gives the ISE index magnitude for a fault contingency, relating all three system arrangements. Table 4.3 shows the improvement in damping performance using SMES units. With the rate of change of frequency as the regulating variable, the eigenvalue profile shows better performance with shifting of the eigenvalues to the left half of the plane. *The aggregate damping ratio value for derivative control is nearly double than that of no SMES system.*

Remark 4.2: Damping performance assessment using local oscillation modes and damping factors is performed in MATLAB® platform using the inbuilt linmod and eig commands.

Table 4.2   ISE scenario for three-phase short-circuit contingency

Topology	ISE
No SMES	119.17
$\Delta\omega$-SMES control	32.48
Derivative-SMES control	**24.663**

Table 4.3   Analytical index: aggregate damping factors

Topology	Sum
No SMES	17.11
$\Delta\omega$-SMES control	22.04
Derivative-SMES control	**39.931**

It is evident from Tables 4.4 and 4.5 that the derivatively controlled SMES strategy curtails the first swings robustly. To stress on the damping as well as the terrific overshoot curtailment imparted by the derivative-based SMES in the wind-embedded WSCC system, a plot of frequency deviation of generator 1 versus the COI frequency can be seen in Figure 4.8. The role of proportionally controlled (GA tuned), PID controlled (GA tuned) and fuzzy logic controlled ([36]) Δf SMES is analysed in light of derivative (or ROCOF)-based SMES [14].

4.5.2   Case B

This section explores the impact of SMES operating in a virtual synchronous generator on transient stability. The test system used in this case study is a modified 18-machine, 70-bus system (Figure 4.18). The 68-bus system consists of five areas with 16 synchronous generators. New England Test System (NETS) and New York Power System (NYPS) are represented by a group of generators, while area 3, area 4 and area 5 are represented by a group of generators (Figure 4.18). Generator 13 also represents a small subarea. Instead of using a singular approach, a distributed approach is adopted, where two SMES units are connected at buses 9 and 10 in NETS and NYPS, respectively. Bus 69 is connected to bus 22 through a transformer. Bus 70 is connected to bus 17 through another transformer. The value of transformer reactance is taken as j0.01. At buses 69 and 70, two doubly fed induction generator (DFIG)-based wind farms each of capacity 87 MW are connected. Bus data, line data and alternator data can be found in Reference 37.

Table 4.4   Absolute peak overshoots for fault contingency

System variables	No. SMES	$\Delta\omega$-based SMES	Derivative SMES control
$\omega_2 - \omega_1$ (p.u)	1.4008	0.8398	**0.5581**
$\omega_3 - \omega_1$ (p.u)	1.7849	0.9018	**0.6123**
P_1 (p.u)	2.4974	1.6316	**0.8061**
P_2 (p.u)	1.4312	1.2998	**1.2737**
P_3 (p.u)	0.7903	0.7632	**0.6102**

Table 4.5   Absolute peak overshoots for line outage contingency

System variables	No. SMES	$\Delta\omega$-based SMES	Derivative SMES control
COI frequency (p.u)	0.001582	0.001441	**0.000931**
P_{wf} (p.u)	0.08894	0.07295	**0.06391**
V_{wf} (p.u)	0.01502	0.01388	**0.01269**

The performance of SMES as VSG against SMES and absence of SMES is validated. Two cases are discussed: a step wind disturbance and a three-phase fault in NETS and NYPS.

Scenario 1: The wind turbines connected to DFIGs at buses 69 and 70 were simultaneously subjected to a step wind disturbance of type shown in Figure 4.9. The frequency deviation of NETS and NYPS is shown in Figure 4.10(a) and (b) respectively. The voltage at buses 69 and 70 is shown in Figure 4.10(c) and (d), respectively, and the following conclusions are drawn.

1. The improvement in frequency deviation by using SMES as VSG is significant in both areas. The maximum frequency deviation in both areas is 0.012 Hz, which is very less compared to a frequency deviation of 0.02 Hz, which occurs when no energy storage is connected, and 0.019 Hz, which occurs when SMES is not operating in VSG mode. The slope displayed in the zoomed-in portion of Figure 4.10(a) and (b) depicts that the ROCOF is the least with SMES-VSG topology.
2. The steady-state error in NETS and NYPS reduces to 0.007 p.u in case when SMES is operating as VSG, while it is 0.01 p.u with SMES and without SMES.
3. The improvement in voltage is clearly visible. The maximum deviation in voltage (p.u) in NETS is 0.9905 with SMES as VSG, which is comprehensively better than 0.9888, and 09887, which occur with and without SMES. The maximum deviation in voltage (p.u) in NYPS is 0.9484 with SMES as VSG, which is

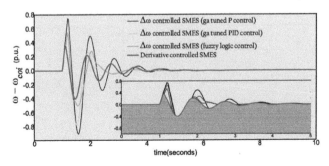

Figure 4.8   A comparative plot on peak frequency overshoot reduction for a 200-ms 3-ϕ fault at bus 6 [14]

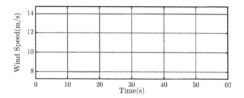

Figure 4.9   Step wind disturbance

comprehensively better than 0.9476 and 0.9475, which occur with and without SMES. Also the peak overshoot at bus 69 after 20 s is significantly reduced with SMES as VSG.

Scenario 2: The test system shown in Figure 4.18 is subjected to a three-phase fault at bus 28 and bus 31 in NETS and NYPS, respectively.

a) Three-phase fault at bus 28

Due to the larger time constant, the mechanical input P_m to the generator remains constant. As a fault occurs in a power system, the electrical power P_e of the synchronous generator reduces. An imbalance of electrical and mechanical power results in an increase in relative rotor angle (δ_{ij}) of the synchronous generator. If the increase in relative rotor angle is large, the generator may lose synchronism. During charging, SMES draws power from the system to charge SMES coil, as a result of which the overall system load increases. As a result, the accelerating power

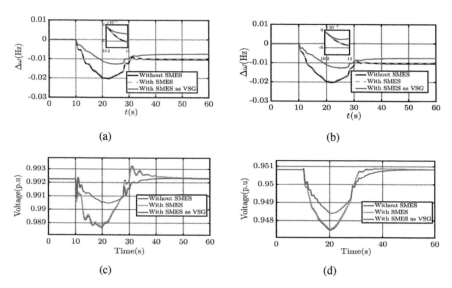

Figure 4.10   Wind disturbance: (a) frequency of NETS, (b) frequency of NYPS, (c) voltage (p.u) at bus 69 and (d) voltage (p.u) at bus 70

decreases, and then (δ_{ij}) also decreases. The overall transient stability of the system increases. Figure 4.11(a) shows that for a fault duration of 142 ms, generator 9 goes out of synchronism when no storage is present. However, it maintains synchronism when SMES is connected. It shows the rotor angle of generator 9, which is very close to bus 28 at which a short circuit is simulated. Figure 4.11(b) shows that for a fault duration of 145 ms, generator 9 remains in synchronism only when SMES is operating in VSG mode. This clearly validates improvement in transient stability with SMES-VSG.

b) Three-phase fault at bus 31

A three-phase short circuit is simulated at bus 31. Figure 4.11(c) shows that for a fault clearing time of 300 ms, generator 10 goes out of synchronism when no storage is present while it maintains its synchronism when SMES is connected and SMES-VSG.

Figure 4.11(d) shows that for a fault clearing time of 305 ms, generator 10 remains in synchronism only when SMES is operating in VSG mode. This clearly validates improvement in transient stability with SMES-VSG.

4.5.3   Case C

This subsection validates the potency of the proposed DPC-equipped SMES along distributed loci as distributed-ESS (DESS) topology and along single nodes as single-ESS (SESS) topologies. Although such faults are a rare occurrence, the

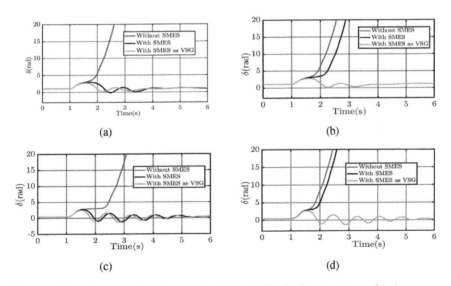

Figure 4.11   Rotor angle of generator 9 for (a) fault clearing time of 142 ms, (b) fault clearing time of 145 ms, rotor angle of generator 10, (c) fault clearing time of 300 ms and (d) fault clearing time of 305 ms

motive of simulating such faults remains to test the transient curves of the system under such intense disturbances. A 3-ϕ short circuit is simulated at $t = 1$ s on the line connecting buses 4 and 5. It is cleared by tripping the defective line in 130 ms. A reoriented form of the exemplar Western System Coordinating Council is chosen as the model system, with an additional DFIG-fed wind farm contributing a share of renewable power (50 MW). A vivid comparison of the abilities of DESS over SESS using DPC-based SMES as the ESS representation can be seen. Tremendous first swing reductions are witnessed in a DESS as compared to a SESS and more so in comparison with a storage-deficient wind-WSCC system multi-area system. The machine node (low inertia constant)-based SESS shows appreciable improvement over that of load bus incorporated SESS, where the latter even shows unstable behaviour when subjected to longer span contingencies with huge peak overshoots.

Results and discussion

1. Considerable frequency regulation is witnessed with DESS topology for both fault situations with quick recovery to pre-fault levels in less than 4 s. Figure 4.12(a) and (b) depicts that the generator-SESS modelled system also shows stability retention in both fault studies like DESS. The average frequency deviation reduction using a DESS amounts to a massive 85–90% *vis-a-vis* the basic system.

2. The transient behaviour in terms of rotor angle oscillations is also plotted, and the average damping improvement of the order of 50–70% is reflected in DESS models. Despite the detrimental contingencies, the rotor angles are strictly maintained at their nominal values in a duration of 4–5 s using DESS-based systems under both fault scenarios. The effective damping imparted by the storage-based power system can be clearly noticed in Figure 4.12(c) and (d), where the fault is cleared in 0.13 s.

3. The role of DESS in improving the voltage profile of the multi-machine system in a fault situation can be witnessed. The DESS shows a roughly 20–30% smooth textured voltage in Figure 4.12(i)–(l) than the storage deficient system. Not only the DESS topology but the SESS topology when incorporated at the low inertia generator bus also shows considerable damping contribution. The smoothing of voltages is seen in generator bus SESS as well as DESS plants, with appreciably more improvement in the latter.

4. Figure 4.12(e)–(h) shows the output power profile of the four machine buses under the 130 ms fault contingency. The combined effect of erratic wind as well as fault is overcome easily with the ESS-clad systems. The power oscillations are also reduced to the order of an average 82–89%.

5. In addition, Figure 4.12(m) conveys that the power handled by the storages lies within the controller power constraints. In fact, the stress on the single storage is prominent, as the power excursions are almost double that of the distributed storage device. *The combined size of the DESS is clearly lesser than both SESS topologies.* However, with all SESS options, the storage power exchange requires a minimum capacity of 80 MW resulting in higher converter ratings.

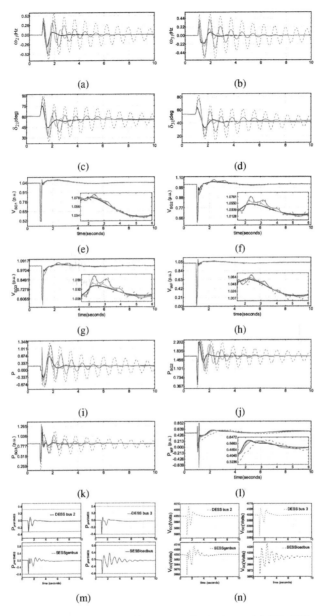

Figure 4.12   *Transient responses: frequency deviations: (a) SG_2-SG_1,*
(b) SG_3-SG_1, rotor angle deviations with respect to SG_1:
(c) SG_2-SG_1, (d) SG_3-SG_1, active power output profile: (e) SG_1,
(f) SG_2, (g) SG_3, (h) WG bus, voltage profile: (i) SG_1, (j) SG_2,
(k) SG_3, (l) WG bus, (m) SMES active power profiles, (n) SMES DC
link voltage profiles; legend—DESSgen (black solid), SESSgen (red
dashed), SESSload (green dotted) and noESS (grey dashed).

The total power capacity of SESS roughly amounts to over 27% of the total system capacity, while DESS only touches about 24% of the same. The inclusion of a DESS hence causes a net reduction in ESS size and cost curtailment in terms of installation capital. Net storage size reduction is achieved using DESS on low-H generator loci, with excursions within 72 MW only. In addition to this, the DPC-based SMES allows the power exchange of the storage to be within the converter power constrained limits. It is evident from Figure 4.12(m) that the power limits (upper as well as lower) are never violated.

6. A similar scenario can be seen with the voltage profile at the DC link. The first swings as well as the oscillations of the voltage are reduced using DESS topology [see Figure 4.12(n)]. The oscillations using SESS topologies at generator bus as well as load bus show at least 10–17% of larger voltage swings. DESS topology maintains a fairly constant DC link voltage status.

4.5.4   Case D

This section explores the deployment of constrained neural-adaptive predictive control equipped DESS units along generator buses and load buses in the considered hybrid 9 bus system (see Figure 4.17). The modelling and control of a 50-MW DFIG-fed wind farm have been performed with dynamic equations detailed in Reference 38. With this aim, time-domain simulations are performed on the study system with DESS incorporation at various nodes. The generator bus-oriented ESS topologies are explored along buses 2 and 3 of the power system. The load buses considered fit for DESS deployment are buses 4 and 6. A three-phase fault is simulated in the line connecting buses 4 and 5. The duration of the fault-on period is 150 ms, after which the fault-ridden line is tripped. The location of DESS placement is particularly chosen along the fault line ends, to impart maximum stability strengthening to the turbulence struck system. Simulation results based on MATLAB®/Simulink® environs validate the efficacy of placement of SMES as generic P-Q controlled devices along generator buses in a multi-machine system.

Remark 4.3: All simulations in this case study are performed with a 100-μs time step, with fourth-order Runge–Kutte mathematics as a model solver.

The following inferences are made from the time-domain simulations.

1. The deployment of energy storages along generator buses reduces the frequency nadir by 13.7–25% [see Figure 4.13(a) and (b)]. Clearly, an average 17–26% more reduction in frequency deviation is achieved using DESS deployment on generator bus nodes than with load bus-based DESS.

2. Excellent overshoot reduction is again achieved in rotor angle oscillation damping with generator bus-based DESS along generator buses is associated with a quicker settling of machine angle oscillations as well as peak overshoot curtailment by 27–33.8%. Comparative plots can be seen in Figure 4.13(c) and (d).

3. Comprehensive power smoothing is witnessed for all electrical power outputs of network machines [see Figure 4.13(e)–(h)]. Significant power smoothing is

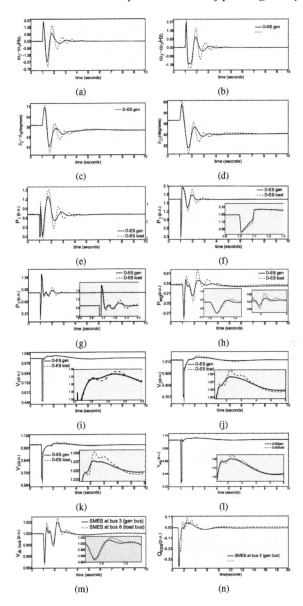

Figure 4.13   *Transient responses: frequency deviations: (a) SG_2-SG_1,*
(b) SG_3-SG_1, rotor angle deviations with respect to SG_1:
(c) SG_2-SG_1, (d) SG_3-SG_1, active power output profile: (e) SG_1,
(f) SG_2, (g) SG_3, (h) WG bus, voltage profile: (i) SG_1, (j) SG_2,
(k) SG_3, (l) WG bus, (m) SMES DC link voltage profiles and
(n) SMES reactive power profiles along generator buses are
associated with a quicker settling of machine angle oscillations as
well as peak overshoot curtailment by 27.

seen on the DFIG-fed WF bus, using SMES on distributed generator bus loci. Even though the load bus-based DESS is chosen specifically along the fault nodes 4 and 6, the DFIG-fed WF performance is far more promising than the other DESS topologies.

4. Voltage ride through and control, post fault application using DESS is evident from all voltage graphs [Figure 4.13(i) and (l)]. The generator bus-based DESS performs better with 16–21% better smoothing and peak overshoot reduction when the system is momentarily hit by a transient blackout.

5. Figure 4.13(m) portrays the behaviour of the DC link voltages of the two SMES units under different placements. It is evident that the SMES unit along bus 6 (load bus) faces vibrant excursions in the DC capacitor voltage. The other SMES scenario, however, is met with intense damping and smoothing of the DC link voltage, as the 6-kV set point voltage is attained within the first 3 s of the contingency.

6. A productive reactive power exchange is facilitated by the ESS units when based on generator nodes, as can be seen in Figure 4.13(n). This is the reason behind a voltage restoration, as the ESS units compensate for the deficient VARs, with minimum Q-power delivery. The exchange of reactive power with SMES placement on bus 3 is approximately half of the exchange when placed on bus 6. In addition, the reactive power exchange oscillations are also appreciably lower.

4.5.5   Case E

This section explores the impact of heuristic fuzzy logic control and parametric fuzzy predictive control schemes of SMES on transient stability. The model system with SMES governed via the aforementioned control strategies is tested by simulating a network fault. The fault location is arbitrarily set at the line linking buses 6–9 of the system. Line disconnection is assumed to rectify the faulty point after 113 ms of fault on period. The following inferences are made from the time-domain simulations:

1. When a system encounters load fluctuations, it is the kinetic energy of the alternator rotors that rescues the system by providing or absorbing the required slack energy. Immediately post rotor dynamic action, the primary and secondary frequency regulator setups come to the forefront by altering the active power set points of the machines. In primary frequency control, the mechanical torque provided by the turbine governors is regulated in accordance with the active power mismatch. This regulation damps the electromechanical oscillations evident in the first tens of seconds. Even though mathematically, a steady-state deviation existing in the system frequency is as good as any new operating steady state, because the alternators might be coherent. However, this deviation ought to be maintained as close to zero as possible, because the overall kinetic energy of the coherently swinging alternators is not the same as before. This is when the

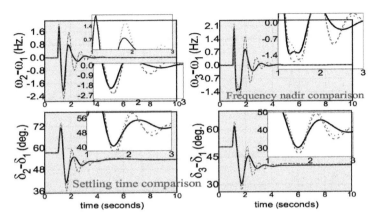

Figure 4.14   Frequency and rotor angle deviation plots; PFPC (black solid), DPC (red dashed) and HFLC (blue dotted)

secondary frequency control (SFC) comes to the forefront. In other words, the kinetic energy of the coherent alternators is driven back to its nominal levels. The time scale of SFC lies over tens of minutes and is generally slow. Because the disturbance is transient, the time scale for stability assessment falls within the first 10 s only. It is at this end; an energy storage offers abundant support.

Clearly, an average 15–22% more reduction in frequency deviation is achieved using PFPC employed SMES than with DPC and HFLC determined SMES action (see Figure 4.14). Significant frequency of nadir curtailment is evident. The incorporation of PFPC-equipped SMES is associated with a quicker settling of machine angle oscillations as well as peak overshoot curtailment by 25–30.5%. The machines are seen to be colloquially in step as no progressive digression of generator angles can be seen. Table 4.6 highlights the significant frequency of nadir curtailment offered by the PFPC determined SMES action. It also makes clear that the transient stability support is effectively much higher using the PFPC strategy, since the rotor angles face minimal excursions.

2. Comprehensive power smoothing is witnessed for all electrical power outputs of network machines (see Figure 4.15). The peak power overshoots are also reduced by employing PFPC control of the generalized energy storage device.

Table 4.6   Frequency nadir and rotor angular deviations

Deviation variable	DPC	HFLC	PFPC
ω_{21}(Hz.)	2.6010	2.592	**2.011**
ω_{31}(Hz.)	1.580	1.589	**1.336**
δ_{21}(∘)	21.839	20.16	**16.94**
δ_{31}(∘)	26.252	22.54	**20.526**

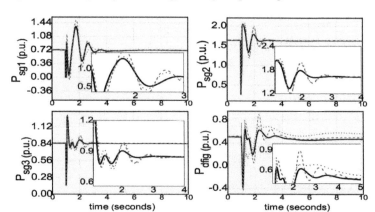

Figure 4.15   Power smoothing evident from electrical power outputs; PFPC (black solid), DPC (red dashed) and HFLC (blue dotted).

The settling time using the proposed technique varies between 38 and 76% of that of the other techniques when applied independently (Table 4.7).

3.   Voltage ride through and control, post fault application using PFPC is evident from all voltage graphs (Figure 4.16). The proposed PFPC-based SMES performs better with 16–21% better smoothing and peak overshoot reduction when the system is momentarily hit by a transient blackout.

4.6   Summary

In this chapter, the proficiency of SMES technology in improving the transient stability of power grids anticipating the intermittent power outputs of wind energy sources is explored. The basic model architecture encompassing the inductor coil and the PCDU is described. The detailed mathematical modelling of various SMES control algorithms, viz. derivative SMES control, SMES contrived as VSG, discrete predictively controlled SMES based on a dynamic constrained window as well as neural network, intelligent fuzzy logic controlled SMES and a selective amalgamation of the said techniques is outlined. The chapter is supplemented by the

Table 4.7   Settling times in seconds from Figure 4.15 studying power swings

Deviation variable	DPC	HFLC	PFPC
P_{sg1} (p.u.)	2.34	2.281	**1.87**
P_{sg2} (p.u.)	1.869	1.872	**1.159**
P_{sg3} (p.u.)	2.79	2.781	**1.085**
P_{dfig} (p.u.)	3.03	-	**2.380**

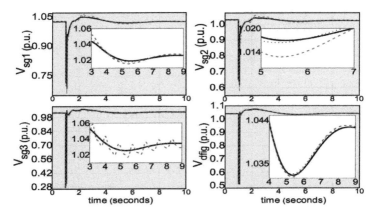

*Figure 4.16   Voltage smoothing profile of machines; PFPC (black solid), DPC
            (red dashed) and HFLC (blue dotted)*

time-domain simulations performed on the WSCC and the 18-machine, 70 bus system for various case studies.

4.7   Appendix

1. A Western System Coordinating Council (WSCC) that emerged around the late 1960s, and was later referred to as the Western Electricity Coordinating Council after the merger of some additional transmission bodies in 2002, is considered the testbed system in this treatise for performing extensive stability studies (see Figure 4.17). The old connotation of the system as WSCC is preferred in this chapter.
2. The 68-bus system is a reduced-order equivalent of the inter-connected NETS and NYPS, with five geographical regions out of which NETS and NYPS are

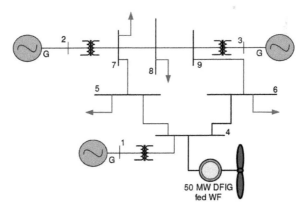

*Figure 4.17   Recast WSCC system—a hybrid conglomerate of alternators and
            wind farm chosen as the testbed system for simulation studies*

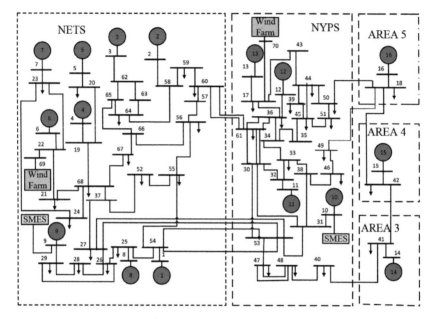

Figure 4.18   Single-line diagram of 18-machine 70-bus system

represented by a group of generators (see Figure 4.18). The power import from each of the three other neighbouring areas is approximated by equivalent generator models. The *dq model* of the 68-bus system is presented on the PES Task Force website on Benchmark Systems for Stability Controls. The reader is directed to Reference 39 for system modelling details.

3.   Renewable power penetration is also addressed simultaneously in the presence of a volatile wind nature. Wind speed modelling is performed in MATLAB®/Simulink® using the autoregressive moving average model. Using this scheme, the wind speed is categorized by three specific components, μ_{wind}, v_{tur} and v_{white} as.

$$v_{wind}(t) = \mu_{wind} + v_{tur} + v_{white} \qquad (4.45)$$

where μ_{wind} represents the average/mean speed of wind, v_{tur} classifies the turbulent wind part and v_{white} is the white noise component.

The turbulent wind component is mathematically a Gaussian noise-defined filter, given as

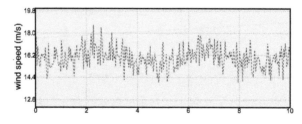

Figure 4.19   Patchy wind speed considered for simulation studies

$$\dot{v}_{tur}(t) = -\frac{1}{T_{gauss}} v_{tur}(t) + \beta_t \qquad (4.46)$$

where, T_{gauss} is a time constant and β_t is an additional white noise shaper.

A highly staggering nature of wind speed is assumed to test the stability of the discussed power system (see Figure 4.19).

References

[1] Ahsan H., Din Mufti M. 'Sweeping power system stabilisation with a parametric fuzzy predictive control of a generalised energy storage device'. *IET Generation, Transmission & Distribution*. 2020, vol. 14(25), pp. 6087–96.

[2] Kiaei I., Lotfifard S. 'Tube-based model predictive control of energy storage systems for enhancing transient stability of power systems'. *IEEE Transactions on Smart Grid*. 2020, vol. 9(6), pp. 6438–47.

[3] Ahsan H., Mufti M., Salam Z., Lone S.A. 'Modeling and simulation of an energy storage based multi-machine power system for transient stability study'. *IEEE Conference on Energy Conversion (CENCON)*; Kuala Lumpur, Malaysia: IEEE, 2009. pp. 78–83.

[4] Iqbal S.J., Mufti M.D., Lone S.A. 'INTELLIGENTLY controlled superconducting magnetic energy storage for improved load frequency control'. *International Journal of Power and Energy Systems*. 2009, vol. 29(4), pp. 241–54.

[5] AhsanH., Mufti M.-D. 'Modeling and simulation of a superconducting magnetic energy storage based multi-machine power system for transient stability study'. *2017 6th International Conference on Computer Applications in Electrical Engineering-Recent Advances (CERA)*; Roorkee: IEEE, 2020. pp. 347–52.

[6] Ahsan H., Mufti M. ud D. 'Distributed storage approach versus singular storage approach: a dynamic stability evaluation'. *International Journal of Power and Energy Systems*. 2019, vol. 39.

[7] Blaabjerg F., Lonel D.M. *Renewable Energy Devices and Systems with Simulations in MATLAB® and ANSYS®*. Boca Raton, FL: CRC Press. 1988. Available from https://www.taylorfrancis.com/books/9781498765831

[8] Jiang H., Zhang C. 'A method of boosting transient stability of wind farm connected power system using S magnetic energy storage unit'. *IEEE Transactions on Applied Superconductivity*. 1988, vol. 29(2), pp. 1–5.

[9] Mitani Y., Tsuji K., Murakami Y. 'Application of superconducting magnet energy storage to improve power system dynamic performance'. *IEEE Transactions on Power Systems*. 1988, vol. 3(4), pp. 1418–25.

[10] Ali M.H., Wu B., Dougal R.A. 'An overview of SMES applications in power and energy systems'. *IEEE Transactions on Sustainable Energy*. 1988, vol. 1(1), pp. 38–47.

[11]   Chen S.-S., Wang L., Lee W.-J., Chen Z. 'Power flow control and damping enhancement of a large wind farm using a superconducting magnetic energy storage unit'. *IET Renewable Power Generation*. 2020, vol. 3(1), p. 23.

[12]   Ali M.H., Tamura J. 'Bin wu'. *IECON 2008 - 34th Annual Conference of IEEE Industrial Electronics Society*; Orlando, FL: IEEE, 2020. pp. 3382–87.

[13]   Xu G., Xu L., Morrow J. 'Power oscillation damping using wind turbines with energy storage systems'. *IET Renewable Power Generation*. 2020, vol. 7(5), pp. 449–57. Available from https://onlinelibrary.wiley.com/toc/17521424/7/5

[14]   Ahsan H., Mufti M.D. 'Comprehensive power system stability improvement with ROCOF controlled SMES'. *Electric Power Components and Systems*. 2020, vol. 48(1–2), pp. 162–73.

[15]   Zargar M.Y., Mufti M.U., Lone S.A. 'Adaptive predictive control of a small capacity SMES unit for improved frequency control of a wind-diesel power system'. *IET Renewable Power Generation*. 2020, vol. 11(14), pp. 1832–40. Available from https://onlinelibrary.wiley.com/toc/17521424/11/14

[16]   Du Z.B., Zhang Y., Liu L., Guan X.H., Ni Y.X., Wu F.F. 'Structure-preserved power-frequency slow dynamics simulation of interconnected AC/DC power systems with AGC consideration'. *IET Generation, Transmission & Distribution*. 2007, vol. 1(6), p. 920.

[17]   Bevrani H., Hiyama T. *Intelligent Automatic Generation Control*. Boca Raton, FL: CRC Press. 2019 Dec. Available from https://www.taylorfrancis.com/books/9781439849545

[18]   Pal B.C., Coonick A.H., Jaimoukha I.M., El-Zobaidi H. 'A linear matrix inequality approach to robust damping control design in power systems with superconducting magnetic energy storage device'. *IEEE Transactions on Power Systems*. 2007, vol. 15(1), pp. 356–62.

[19]   Hartmann B., Vokony I., Táczi I. 'Effects of decreasing synchronous inertia on power system dynamics—overview of recent experiences and marketisation of services'. *International Transactions on Electrical Energy Systems*. 2019, vol. 29(12), p. 12. Available from https://onlinelibrary.wiley.com/toc/20507038/29/12

[20]   Kumar A.W., Mufti M.D., Zargar M.Y. 'Dynamic performance enhancement of wind penetrated power system using SMES as virtual synchronous generator'. *International Journal of Power and Energy Systems*. 2019, vol. 41(2), pp. 1–8.

[21]   Naik P.K., Nair N.K.C., Swain A.K. 'Impact of reduced inertia on transient stability of networks with asynchronous generation'. *International Transactions on Electrical Energy Systems*. 2006, vol. 26(1), pp. 175–91. Available from http://doi.wiley.com/10.1002/etep.v26.1

[22]   Driesen J., Visscher K. 'Virtual synchronous generators'. *IEEE Power and Energy Society 2008 General Meeting: Conversion and Delivery of Electrical Energy in the 21st Century, PES*; 2008. pp. 1–3.

[23]   Morren J., de Haan S.W.H., Kling W.L., Ferreira J.A. 'Wind turbines emulating inertia and supporting primary frequency control'. *IEEE Transactions on Power Systems*. 2006, vol. 21(1), pp. 433–34.

[24] Zargar M.Y., Lone S.A., Mufti M.U.-D. 'MATLAB/simulink-based modelling and performance assessment of wind–diesel energy storage system'. *Wind Engineering*. 2006, vol. 42(3), pp. 194–208.

[25] Milano F., Ortega Á. 'Frequency divider'. *IEEE Transactions on Power Systems*. 2017, vol. 32(2), pp. 1493–501.

[26] Ghadimi N., Akbarimajd A., Shayeghi H., Abedinia O. 'Two stage forecast engine with feature selection technique and improved meta-heuristic algorithm for electricity load forecasting'. *Energy*. 2018, vol. 161, pp. 130–42.

[27] Phillips C.L., Nagle H.T. *Digital Control System Analysis and Design*. Hoboken, NJ: Prentice Hall Press; 2007.

[28] Syed A., Mufti M.U.D. 'PREDICTION-based adaptive control of SMES for multi-area power systems'. *International Journal of Power and Energy Systems*. 2021, vol. 41(4), p. 4.

[29] Patyra M.J., Mlynek D.J. *Fuzzy Logic: Implementation and Applications*. Singapore: Springer Science & Business Media; 2012.

[30] Wang S., Tang Y., Shi J, *et al.* 'Design and advanced control strategies of a hybrid energy storage system for the grid integration of wind power generations'. *IET Renewable Power Generation*. 2015, vol. 9(2), pp. 89–98.

[31] Hasanien H.M., Muyeen S.M. 'Design optimization of controller parameters used in variable speed wind energy conversion system by genetic algorithms'. *IEEE Transactions on Sustainable Energy*. 1988, vol. 3(2), pp. 200–08.

[32] Ahsan H., Mufti M.D. 'Modeling and control of a generalized energy storage device for stabilizing frequency and voltage oscillations'. *2019 9th International Conference on Power and Energy Systems (ICPES)*; Perth, Australia: IEEE, 1988. pp. 1–6. Available from https://ieeexplore.ieee.org/xpl/mostRecentIssue.jsp?punumber=9098828

[33] Goldberg D.E., Holland J.H. 'Genetic algorithms and machine learning'. *Machine Learning*. 1988, vol. 3(2/3), pp. 95–99.

[34] Ahsan H., Mufti M.D. 'Systematic development and application of a fuzzy logic equipped generic energy storage system for dynamic stability reinforcement'. *International Journal of Energy Research*. 1988, vol. 44(11), pp. 8974–87. Available from https://onlinelibrary.wiley.com/toc/1099114x/44/11

[35] Milano F. *'Power System Modelling and Scripting' in Power System Modelling and Scripting*. Berlin, Heidelberg: Springer Science & Business Media; 1988.

[36] Chaiyatham T., Ngamroo I. 'Improvement of power system transient stability by pv farm with fuzzy gain scheduling of PID controller'. *IEEE Systems Journal*. 2020, vol. 11(3), pp. 1684–91.

[37] Pal B., Chaudhuri B. *Robust Control in Power Systems*. New york, USA: Springer Science & Business Media; 2006.

[38] Ahsan H., Mufti M.-D. 'Modelling and stability investigations of an aggregate wind farm–fed multi-machine system'. *Wind Engineering*. 2020, vol. 44(6), pp. 596–609.

[39] Kumar A.W., din M.M., Zargar M.Y. 'Adaptive predictive control of flywheel storage for transient stability enhancement of a wind penetrated power system'. *International Journal of Energy Research*. 2020, vol. 46(5), pp. 6654–71. Available from https://onlinelibrary.wiley.com/toc/1099114x/46/5

Chapter 5

Enhancement of load frequency control in interconnected microgrids by SMES

Sayed M. Said[1], Emad A. Mohamed[1], Mokhtar Aly[2], and Emad M. Ahmed[1, 3]

The increased penetration levels of various renewable energy sources with their stochastic nature have increased interest in the incorporation of energy storage systems (ESSs) for load frequency control function in isolated and/or interconnected electrical power systems. Among the existing ESSs technologies, the superconducting magnetic energy storage (SMES) has become favorable in several applications compared to the other existing ESSs. This chapter introduces the problem of frequency regulation in interconnected microgrid (MG) systems with their modeling for two-area power systems as a case study. Then, the various utilized devices and control methods in the interconnected MG for frequency regulation are introduced. Finally, a case study of some of the widely employed controllers is introduced. The various characteristics of renewable energy, loading, and SMES are considered in the presented results.

5.1 Introduction

Recently, modern power systems have been equipped with several renewable energy sources (RESs), which can cause more and more challenges to electrical power systems [1]. By increasing the penetration levels of RESs into the power systems, the total system inertia will be decreased, especially during the presence of solar power generation systems. A low inertia level can lead to the power system dramatically reaching the frequency instability region, where there is a significant difference between the total input and total output power of the electrical system [2, 3].

Figure 5.1 shows the complete structure of the interconnected microgrid (MG) system, which consists of multi-MG systems connected via an electrical tie-line.

[1]Department of Electrical Engineering, Aswan University, Aswan 81542, Egypt
[2]Facultad de Ingeniería y Tecnología, Universidad San Sebastián, Bellavista 7, Santiago, Chile
[3]Department of Electrical Engineering, Jouf University, College of Engineering, Sakaka 2014, Saudi Arabia

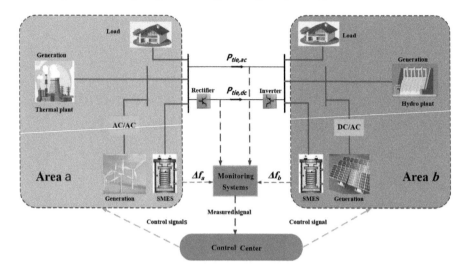

Figure 5.1   Structure of the single line diagram for interconnected two-area electrical power system

Each single MG system comprises conventional power plants (i.e., thermal and hydraulic power plants), RESs (such as wind energy and solar energy), different loads (i.e., domestic, industrial, and commercial loads), and energy storage systems (ESSs), additionally the control center and protection systems.

Frequency control in power systems is usually classified into three main control levels. These control levels are the primary control, the secondary control (i.e., load frequency control (LFC)), and the tertiary control as shown in Figure 5.2. The aim of all control actions is to maintain the desired megawatt output power of a generator matching the changing load, to assist in controlling the frequency of larger interconnection, and to keep the net interchange power between pool members at the predetermined values.

The primary control is implemented through governor control installed in each generating unit that starts within seconds of a disturbance. Its main role is to stabilize the frequency system by activating the reserve units to compensate for the requested load demand for increasing the system frequency but cannot restore the frequency to its nominal value. The primary control can solve the problem only during the first 30 seconds after occurring of disturbance. While the secondary control is used for the frequency fluctuation of a period of about 30 minutes. It is responsible for adjusting the frequency in the electric power systems considering two major goals: (1) preserving the frequency value within an acceptable range and (2) regulating the interchange of the tie-line power between the multi-areas interconnected power systems. The tertiary control refers to the economic dispatching control of units (i.e., the manual activation of power reserves by the system operator). Its actions on a 30-minute-to-hours time scale.

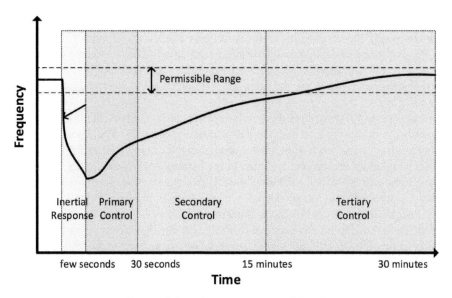

Figure 5.2   Frequency control levels

The employment of power electronics converter in the integration of RESs with the utility grid is necessary for improving their power quality, maximizing the power extraction, and increasing the energy efficiency [4, 5]. However, the increased levels of penetrations of RESs and their associated power electronics converters have led to a significant decrease in the overall inertia of electrical power systems [6, 7]. The reduced system inertia represents the main cause of voltage and/or frequency fluctuations, power system instability, and harmful consequences in electrical power systems. Therefore, presenting adequate control systems and efficient disturbance mitigation solutions is essential for preserving secure and stable power system operation [8].

The virtual synchronous generator (VSG) methods have found wide applications in mitigating the effects of low inertia problems resulting from the high penetration levels of RESs in electrical power systems [9]. The main idea of VSG controllers is to emulate the performance and operation of classical synchronous generating stations without the need for prime movers. In the literature, VSG-based solutions have proven improved stability and performance in electrical power systems. Additionally, automatic load frequency control (ALFC) techniques have also been widely employed for achieving stable operation and adequate control of the frequency and voltage of power systems during the sudden load variations, and the imbalance between the generation and loading [10]. The ALFC techniques fail at mitigating the fast transients in frequency of electrical power systems [11].

On another side, virtual inertia control (VIC) techniques have been utilized in combination with the ALFC techniques in improving the performance of frequency regulation of electrical power systems [12, 13]. The VIC techniques have achieved superior performance with various power system structures [14, 15]. In

VIC techniques, ESSs are employed for regulating the power system frequency and for mitigating the various disturbances through emulating the inertial response of VSG. The superconducting magnetic energy storage (SMES) has found wide applications in VIC compared to the other existing ESSs [16]. The fast response and high power density of SMES are the main key elements behind their wide applications in VIC [16].

LFC performs a significant role in preserving the stability of the interconnected power systems during load disturbances and RESs variations. The deviations of the nominal frequency and tie-line power are being the driving signals for the LFC in single- and multi-area power systems. In the literature, different control techniques have been addressed for LFC based on integral- and fractional-order (FO) controllers (e.g., proportional-integral-derivative (PID), fractional-order PID (FOPID), and tilt integral derivative with filter), modern controllers (e.g., sliding mode control (SMC) and model predictive control (MPC)), and intelligent controllers (e.g., fuzzy logic control (FLC), artificial neural network (ANN), and expert systems).

Due to the increased controller parameters, particularly in multi-area power systems, the design of LFC is a cumbersome issue with the increased controller parameters; therefore, the applications of the soft-computing techniques, e.g., particle swarm optimization (PSO), manta ray foraging optimization (MRFO), and AEO, have found a productive area for designing the controller parameters.

Based on the above background, this chapter discusses the frequency regulation issue in the interconnected MG power systems, by taking the two-area MG system as a case study. Several devices and control approaches to address the frequency regulation issue in the interconnected MG system are offered. The behavior of the RESs, ESS, and various load types is considered.

5.2   LFC issues in interconnected MG

5.2.1   Modeling of MG system

The block diagram of the multi-MG system is presented in Figure 5.3. The multi-MG system consists of conventional main generation, wind power generation, solar power generation system, ESSs, and load demand. The load demand is classified into residential, commercial, and industrial loads, and is mainly supplied from the RESs [17]. The conventional generation is modeled as a thermal power plant as follows [16]:

$$\Delta P_m = \Delta P_g \frac{1}{T_g s + 1} \tag{5.1}$$

$$\Delta P_g = \frac{P_n}{T s + 1} * (\frac{-1}{R} * \Delta f - ACE) \tag{5.2}$$

The wind and solar power generations are considered RESs in the selected case study and can be modeled traditionally by the first-order transfer function of a unity gain and time constants T_{WT} and T_{PV}, respectively, as follows [18]:

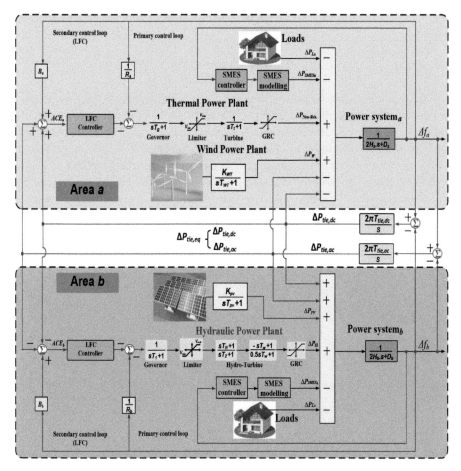

Figure 5.3   Dynamic model of the studied interconnected microgrid two-area electrical power system

$$\Delta P_W = \frac{1}{sT_{WT} + 1} * \Delta P_{Wind} \qquad (5.3)$$

$$\Delta P_{PV} = \frac{1}{sT_{PV} + 1} * \Delta P_{Solar} \qquad (5.4)$$

The ESS can be modeled in terms of the frequency deviation (Δf) as follows:

$$\Delta P_{ESS} = \frac{K_{ESS}}{sT_{ESS} + 1} * \Delta f \qquad (5.5)$$

where H, D, ΔP_m, ΔP_{ESS}, ACE, R, P_n, ΔP_g, and T_g are system inertia, damping system coefficient, change in mechanical power, ESS variation, area control error signal, regulating system frequency, governor speed regulation of the thermal plant, nominal rated power output for thermal plant, governor power deviation of the thermal power plant, and governor time constant of the thermal power plant, respectively.

5.2.2 Formulation of frequency deviation

Complete system modeling is essential for designing the control systems for interconnected power systems. The models of the various existing elements in the electrical power system are combined to construct the overall model. The most appropriate way to construct the model is through using the widely used state space modeling. In the constructed state space mode, x, y, ω, and u are the vectors representing state variables, output states, disturbances, and other control variables, respectively. In the constructed model, the control variables that are considered include the ACE signals in areas (ACE_a and ACE_b) and SMES output power difference in areas (ΔP_{SMESa} and ΔP_{SMESb}). Whereas, A, B_1, B_2, and C are parameter matrices that correspond to linearized state-space modeling of studied system. The mathematical modeling of the state space and its parameters can be expressed as follows:

$$\dot{x} = Ax + B_1\omega + B_2u \tag{5.6}$$

$$y = Cx \tag{5.7}$$

where

$$x = \begin{bmatrix} \Delta f_a & \Delta P_{ga} & \Delta P_{ga1} & \Delta P_{WT} & \Delta f_b & \Delta P_{gb} & \Delta P_{gb1} & \Delta P_{gb2} & \Delta P_{PV} & \Delta P_{ti} \end{bmatrix} \tag{5.8}$$

$$\omega = \begin{bmatrix} \Delta P_{la} & P_{WT} & \Delta P_{lb} & P_{PV} \end{bmatrix}^T \tag{5.9}$$

$$u = \begin{bmatrix} ACE_a & \Delta P_{SMESa} & ACE_b & \Delta P_{SMESb} \end{bmatrix}^T \tag{5.10}$$

$$A = \begin{bmatrix}
-\frac{D_a}{2H_a} & \frac{1}{2H_a} & 0 & \frac{1}{2H_a} & 0 & 0 & 0 & 0 & 0 & -\frac{1}{2H_a} \\
0 & -\frac{1}{T_t} & \frac{1}{T_t} & 0 & 0 & 0 & 0 & 0 & 0 & 0 \\
-\frac{1}{R_a T_g} & 0 & -\frac{1}{T_g} & 0 & 0 & 0 & 0 & 0 & 0 & 0 \\
0 & 0 & 0 & -\frac{1}{T_{WT}} & 0 & 0 & 0 & 0 & 0 & 0 \\
0 & 0 & 0 & 0 & -\frac{D_b}{2H_b} & \frac{1}{2H_b} & 0 & 0 & \frac{1}{2H_b} & \frac{1}{2H_b} \\
0 & 0 & 0 & 0 & \frac{2T_R}{R_b T_1 T_2} & -\frac{2}{T_w} & \frac{2T_2+2T_w}{T_2 T_w} & \frac{2T_R-2T_1}{T_1 T_2} & 0 & 0 \\
0 & 0 & 0 & 0 & -\frac{T_R}{R_b T_1 T_2} & 0 & -\frac{1}{T_2} & \frac{T_1-T_R}{T_1 T_2} & 0 & 0 \\
0 & 0 & 0 & 0 & -\frac{1}{R_b T_1} & 0 & 0 & -\frac{1}{T_1} & 0 & 0 \\
0 & 0 & 0 & 0 & 0 & 0 & 0 & 0 & -\frac{1}{T_{PV}} & 0 \\
2\pi T_{tie,eq} & 0 & 0 & 0 & -2\pi T_{tie,eq} & 0 & 0 & 0 & 0 & 0
\end{bmatrix} \tag{5.11}$$

$$
B_1 = \begin{bmatrix}
-\frac{1}{2H_a} & 0 & 0 & 0 \\
0 & 0 & 0 & 0 \\
0 & 0 & 0 & 0 \\
0 & \frac{K_{WT}}{T_{WT}} & 0 & 0 \\
0 & 0 & -\frac{1}{2H_b} & 0 \\
0 & 0 & 0 & 0 \\
0 & 0 & 0 & 0 \\
0 & 0 & 0 & 0 \\
0 & 0 & 0 & \frac{K_{PV}}{T_{PV}} \\
0 & 0 & 0 & 0
\end{bmatrix}, \text{ and } \quad
B_2 = \begin{bmatrix}
0 & -\frac{1}{2H_a} & 0 & 0 \\
0 & 0 & 0 & 0 \\
-\frac{1}{T_g} & 0 & 0 & 0 \\
0 & 0 & 0 & 0 \\
0 & 0 & 0 & -\frac{1}{2H_b} \\
0 & 0 & \frac{2T_R}{T_1 T_2} & 0 \\
0 & 0 & -\frac{T_R}{T_1 T_2} & 0 \\
0 & 0 & -\frac{1}{T_1} & 0 \\
0 & 0 & 0 & 0 \\
0 & 0 & 0 & 0
\end{bmatrix}
\tag{5.12}
$$

$$
C = \begin{bmatrix}
1 & 0 & 0 & 0 & 0 & 0 & 0 & 0 & 0 & 0 \\
B_a & 0 & 0 & 0 & 0 & 0 & 0 & 0 & 0 & 1 \\
0 & 0 & 0 & 0 & 1 & 0 & 0 & 0 & 0 & 0 \\
0 & 0 & 0 & 0 & B_b & 0 & 0 & 0 & 0 & -1
\end{bmatrix}
\tag{5.13}
$$

5.2.3   Modeling of the SMES system

The configuration of the SMES system as one of the ESSs is presented in Figure 5.4. It comprises a three-phase coupling transformer, voltage source converter (VSC) using an insulated-gate bipolar transistor, and Direct current (DC) link capacitor. The important part of the SMES system is the superconducting coil (SC), which is interconnected to the alternating current (AC) system through a power conditioning system

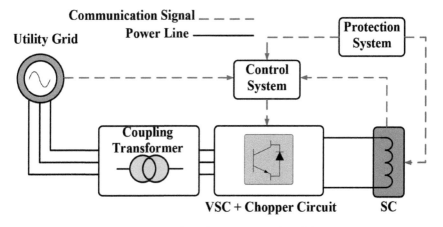

Figure 5.4   The detailed configuration model of the SMES system

(PCS), which includes a converter and DC–DC chopper circuit. The SC is charged from the grid during the normal operation to a set value. After charging the SC, SMES can operate in standby mode without charging or discharging energy. During SMES operation, the SC should be immersed in liquid helium or liquid nitrogen in an insulated vacuum cryostat to maintain it in the superconducting state. Whenever there is a fluctuation in the power system during a disturbance, the discharge starts immediately through the PCS to the AC grid. The control system can restore the power system to an equilibrium condition and the SC charges again to its steady-state value. The complete equations of the SMES system can be summarized as follows:

$$E_{smo} = \frac{1}{2} * I_{smo}^2 * L_{sm} \tag{5.14}$$

$$V_{sm} = (2D - 1) * V_{DC} \tag{5.15}$$

$$P_{sm} = V_{sm} * I_{sm} \tag{5.16}$$

$$E_{sm} = E_{smo} + \int_0^t (P_{sm}dt) \tag{5.17}$$

where E_{smo}, E_{sm}, I_{smo}, I_{sm}, L_{sm}, V_{sm}, P_{sm}, and V_{DC} are the initial SMES energy, current SMES stored energy, initial SMES coil current, updated SMES current, SMES coil inductance, SMES coil voltage, output SMES active power, and voltage across DC capacitor linked between SMES chopper circuit and SMES VSC, respectively.

5.2.4   Stochastic property of RESs and sudden load changes

The natural property of RESs has affected the frequency stability of the multi-MG system. The random variations of wind speed and unexpected changes in solar

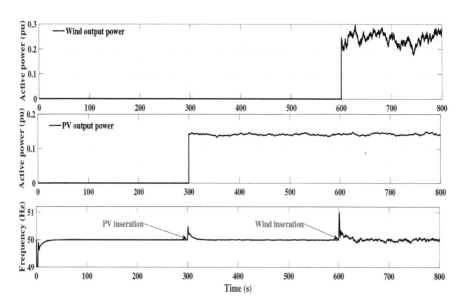

Figure 5.5   Impact of RESs natural properties on system frequency

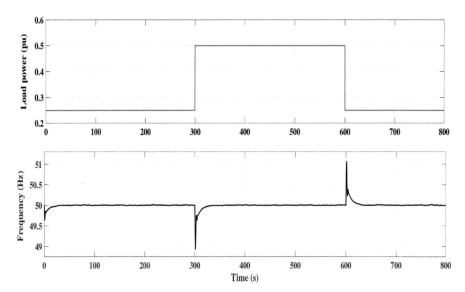

Figure 5.6   Impact of the sudden load change on system frequency

irradiance have a negative response to the frequency stability of the interconnected MG systems. On the other hand, the frequency stability can be affected by the sudden changes in the load demand. With increasing/decreasing load power, the system frequency will change with falling/skipping from the nominal value, respectively.

To clarify the natural property of RESs and the sudden load change, the MG system presented by Said *et al.* [19] is examined to study the response of the MG frequency. MATLAB Simulink package is used to implement the MG system. Figures 5.5 and 5.6 present the behavior of the system frequency during the natural variation of the RESs in addition to the step change in the load power demand. The impact of RESs natural properties is described in Figure 5.5. The negative effect of the RESs properties on the system frequency can be seen especially during the insertion time. Figure 5.6 discusses the impact of the load step change; the system frequency dramatically dropped during the insertion of sudden loads. On the other hand, at the load rejection, suddenly, the frequency instantly goes up over its nominal value. This, in turn, makes the MG system near an unstable region.

5.2.5   Impact of system inertia reduction

The frequency stability can be adversely affected by decreasing the system inertia (H). This reduction results from increasing the penetration levels of RESs, which are interconnected to the MG system. Usually, RESs have no, or small rotating mass compared with the conventional sources, which can be led to a clear difference between total electrical and mechanical power. This difference has the main role in reaching the MG system to the unstable region. Figure 5.7 shows the impact of reduction of MG system inertia on the frequency stability. It can be seen that, by reducing the system inertia, the frequency oscillation dramatically increased until the MG system became unstable as the system frequency dropped to zero, which is shown in Figure 5.7.

5.3   Minimization of frequency and tie-line power oscillations in conventional source-based interconnected MG

5.3.1   Using conventional generation systems

Different periodic components accumulate the change in load or frequency fluctuations of an electrical power system. The balancing between power generation and load demand is maintained in response to load changes by regulating the output of thermal or hydraulic power generators, and the system frequency is kept within a certain range. The inertial response of load and synchronous generator is in charge of regulating short-term load fluctuations of less than 10 seconds. However, load fluctuations lasting more than 10 seconds are controlled by using a turbine governor. Fluctuations ranging from 10 seconds to more than 10 minutes are managed by the LFC, which is linked to a load dispatch center. Moreover, for fluctuations, cycles longer than 10 minutes can be predicted with some accuracy, and the load demands are distributed as economic load dispatching control demand, from the load dispatch center to each plant.

5.3.1.1   Hydraulic power generator

The power change and structure of a hydro turbine are affected by the height of the waterfall. The characteristics of a hydro turbine are extremely complex. It varies according to the operating conditions. The water pressure response in a hydro turbine is initially opposite to the gate position change and then recovers after the transient response. In hydropower plants, generating units are normally adjusted

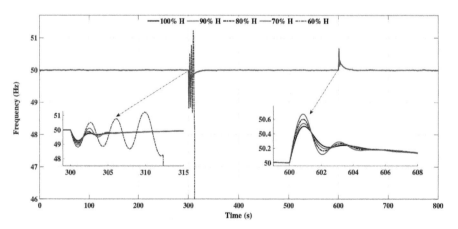

Figure 5.7   Impact of MG system inertia (H) on the frequency stability

for the LFC because the response is faster to raise/lower the power. Because of the inertia of water, hydro turbines are non-minimum phase units.

5.3.1.2   Thermal power generator

The turbine governor in a thermal power plant regulates the flow of steam into the turbine based on the amount of power generated. When demand increases or decreases, the governor detects the unbalanced condition between generation and demand indirectly through changes in frequency. The governor changes the inlet valve position of the steam that enters the turbine based on this frequency value. In this manner, the speed governor balances generation and demand. The steam at high pressure and low temperature is withdrawn from the turbine at an intermediate stage in such turbines. It is returned to the boiler for superheating before being reintroduced at low pressure and high temperature into the turbine. The overall thermal efficiency is increased in this manner. As a result, the reheat turbine transfer function is defined by two-time constants.

5.3.2   Using high-voltage direct current (HVDC) systems

The common AC transmission systems impose some kind of limitation on the transfer of power in the electrical power system, particularly when transmitting between unsynchronized AC systems. The Ferranti effect in long transmission lines, the capacitive effect of underground cables, the influence of line inductance, and the issue of interconnected power system stability are all limitations of AC transmission systems [20–22]. The problem of reactive power due to high-voltage AC (HVAC) cable charging is very important in long transmission lines based on HVAC cables. Furthermore, the length of these cables limits the ability of active power transmission, which decreases as the line length increases. On the other hand, the HVDC transmission system has capabilities that can assist in providing solutions to these existing problems and improving the frequency regulation of electrical power systems, thereby improving the overall stability of interconnected networks. This is possible because the power flows through an HVDC link that can be controlled independently by the phase angle between the source and the load. It can stabilize a network against power fluctuations and allow power transmission between unsynchronized AC transmission systems. Furthermore, it allows power exchange between incompatible networks improves the stability and economy of each grid. It uses an HVDC link to implement a frequency stabilization concept in interconnected power systems, which has been reported in many projects, such as the study in References 23 and 24 demonstrated the benefits of parallel hybrid AC/DC interconnections in improving the dynamics of the frequency controller and dampening post-fault frequency oscillations. The proposed approach in Thottungal *et al.* [25] investigated the effects of an HVDC link in parallel with an HVAC link on the LFC problem for a multi-area power system, as illustrated in Figure 5.8 of the schematic outline of the interconnected areas of an electrical power system using a hybrid HVAC/HVDC tie-line for exchanging power between the two areas while taking system parameter variations into account.

5.3.3   Using flexible AC transmission systems (FACTS) devices

In the last two or three decenniums, the usage of FACTS devices has been widespread in power systems in order to fully utilize existing transmission capacities rather than building new lines. FACTS have an effective technique for managing the tie-line power of an interconnected power system. It injects a changeable series voltage to impact the power flow by varying the phase angle. They control the active power flow in power systems, and due to their high speed, it is well suited for improving power system operation and control. By adding FACTS such as supercapacitor/ultracapacitor, power fluctuations can be properly damped by storing electrical energy during excess generation and deliver power during peak load demand periods within a short time duration. Therefore, these devices play an important role in controlling and managing the active power flow in the tie-line in interconnected power systems for LFC problems. Several FACTS devices have been developed and suggested for LFC, including the static VAR compensator, which is applied as a novel LFC in Reference 26. Moreover, the thyristor control phase shifter (TCPS) is applied to the tie-line power flow of the proposed primary integral controller as an LFC, an interconnected power system [27]. However, a TCPS is developed for providing system frequency regulation of multi-area power systems in coordination with H-infinity and LMI control techniques [28]. Furthermore, the researchers present a coordinated control between TCPS and SMES in the LFC loop with the primary integral controller and all controller and TCPS/SMES parameters are being optimized by PSO [29]. A robust decentralized frequency stabilizer design using static synchronous compensators by taking system uncertainties into consideration is developed [30]. A nonlinear approach is presented based on the coordinated controller of static synchronous series compensator and power system stabilizer by Bhongade *et al.* [31]. These previous approaches of FACTS devices in LFC problem reveal that the improvements of the power system dynamics not only depend on various controllers but also on the addition of FACTS and energy storage devices and utilization of intelligent techniques for optimizing all the power system parameters.

5.4   Minimization of frequency and tie-line power oscillations in RES-based interconnected MG

5.4.1   Using photovoltaic (PV) farms

The PV generation stations can also be employed for the frequency regulation in interconnected MG [32]. The PV generation plants can participate in the frequency regulation by using the deloading technique and ESSs [33]. In the deloading technique, the PV generation plants are operated below their maximum generation power points. By doing that, a certain amount of the remaining active power is employed for supplying the necessary active power to the utility grid during the frequency transients [34]. In conventional operation of PV plants, they are operated continuously at the maximum available output power through employing the widely known maximum power point tracking (MPPT) techniques [35]. The MPPT control

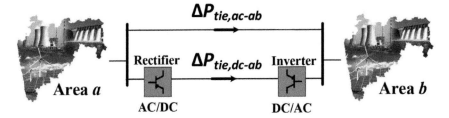

$\Delta P_{tie,ac-ab}$

Rectifier $\Delta P_{tie,dc-ab}$ Inverter

Area *a*

AC/DC   DC/AC

Area *b*

Figure 5.8   Equivalent two area model with connection of AC/DC tie-line

has to continuously operate the PV systems at MPPT according to the variations in the solar irradiance and/or ambient temperature. Whereas, through the derating technique, curtailment is made through operating PV power plants at lower power point than MPPT, which is the deloaded power point [36]. Thence, PV power plants have the ability to support the power system frequency through using the available reserve of the difference between the power at MPPT and the power at deloaded power point. The main operation of the derating technique is shown in Figure 5.9. It can be seen that the operating power of PV power plants under this technique will be usually lower than the power at MPPT. Additionally, it can be seen that there are two different operating regions at the same output power level with one on the right side and the other on the left side of the original MPPT. The PV power plants are operated by the point in the right side of MPPT that corresponds to the higher operating PV voltage for preserving stable operation of PV plants [37].

On another side, various ESSs can be employed with PV power plants for achieving the frequency regulation function. In these methods, the PV generation and the ESSs are integrated together with the power inverters [38]. The ESSs can provide the additional required active power in order to mitigate the various disturbances. Several types of ESSs are used in the literature for this function, such as the commonly used battery ESSs, supercapacitors ESSs, and flywheel ESSs.

5.4.2   Using wind plants

The wind power plants can also be controlled for improving the frequency response of interconnected power systems [39]. There are three main ways to make the contribution of wind power plants in the frequency regulation: through the use of ESSs, the employment of deloading techniques, and providing an inertial response. For the use of ESSs, it is the same as in PV power plants, wherein the energy storage devices are utilized for proving the required active power during the various disturbances [40].

Whereas, the derating techniques are made through the pitch angle control methods or through the over-speed control methods. In the pitch angle control method, the rotor speed is maintained at the maximum power point by increasing the operating pitch angle for constant wind speed. By doing that, the supplied active power from the wind plant is reduced to a value below the maximum available active power [41]. Thence, a certain portion of the reserve active power is available for the purpose of additional electrical power supply during the frequency variation conditions.

Whereas, for the over-speed control method, the deloaded active power is shifted to the right side of the MPPT operating point while preserving the pitch angle for constant wind speed value. During the frequency support operation, a reduction of the rotor speed is needed for releasing the kinetic energy to support the system [42].

On another side, the participation of wind power plants in the frequency regulation through providing the inertial response is made by increasing the outputted active power from wind for few seconds [32]. Additional control loops are added to the basic active power controller and they are only activated during disturbances. The inertial response can be made through the droop control, the emulation of hidden inertia, and the fat power reserve control [43]. In droop control methods, the emulation of governor behavior in the synchronous generators is made according to the frequency changes. In which, the supplied active power by wind generation is changed proportionally to deviations of the frequency. Whereas, the emulation of hidden inertia is based on emulating the behavior of conventional synchronous generators. There are two main ways to implement this type of control: through using a single loop or using two-loop control. In the fast power reserve method, the stored kinetic energy in rotating masses of the wind turbine is supplied to the power grid in the form of additional active powers, which are supplied in the case when frequency deviation exceeds predefined limits [44].

5.4.3   Using electric vehicles

With the increased concerns about reducing carbon emissions to face climate changes, green technologies have found wide utilization in various sectors. The electrical vehicles (EVs) represent an efficient replacement of fuel-based vehicles in small-, medium-, and high-scale vehicles. The increased installations of EVs in electrical systems have opened the way to add more functionalities to them to participate in the regulation of electrical power systems. The vehicle-to-grid (V2G) concept has become more popular in recent years [45]. In V2G systems, the inherent batteries of EVs are employed for decoupling the system generation and the load demand, particularly during the reduced RESs generation profiles under their expected output power [46]. The charging/discharging operations of EVs batteries are controlled through their bidirectional power electronics converters for supporting the grid, adjusting the grid operating frequency, mitigating the load variations, and minimizing the tie-line power oscillations [47]. In recent research, the design of various controllers has found researchers' interests in considering the EV participation, the characteristics of RESs and their high penetration levels, and the nonlinearities of generation rate constraint (GRC) and governor dead band [48].

Several modeling methods have been presented in the literature to represent the EVs for LFC purposes. The model by Ota *et al.* [49] has found wide applications in LFC applications with counting for internal characteristics of EVs. The dynamic modeling from Falahati *et al.* and Luo *et al.* [50, 51] for EV batteries, which is based on Thevenin equivalent modeling of EVs, is shown in Figure 5.10. The model includes the open-circuit voltage source V_{oc} that is the function of the initial state of charge (SOC) of EV batteries. The source is connected with the series resistance

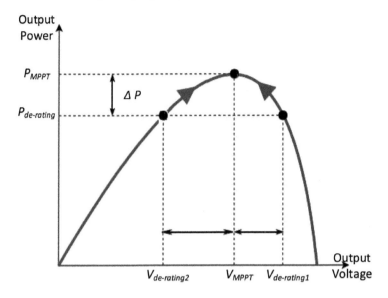

Figure 5.9   The main operation of the derating technique with PV power plants

R_s and the parallel RC circuit (with R_t, C_t) that represent the transient overvoltage effects. The output terminal voltage of EVs can be obtained by combining V_{oc} with the series and parallel elements as represented in Figure 5.10. The Nernst equation can be used for implementing the relation among V_{oc} and EVs SOC as follows (5.18), [49]:

$$V_{oc}(SOC) = V_{nom} + S \frac{RT}{F} ln \left(\frac{SOC}{C_{nom}-SOC}\right)$$ (5.18)

where V_{nom} denotes the nominal voltage of EVs and C_{nom} represents the nominal EV capacity in AH. Whereas, S denotes the sensitivity parameter among V_{oc} and the EV SOC. Moreover, F, T, and R are the Faraday constant, the operating temperature, and the gas constant, respectively.

5.4.4   Using energy storage devices

The various ESSs can be incorporated for reducing the frequency deviations to their minimum values [52]. The battery ESSs and SMES devices are widely employed for frequency regulation in electrical power systems. In the study by Aditya and Das [53], battery ESSs have been utilized for supplying the sudden real power requirements and hence they are effective at mitigating peak frequency in addition to tie-line power deviations. Additionally, battery ESSs have been applied for improving the dynamics of LFC in electrical supply systems of West Berlin by Kunisch *et al.* [54]. Additionally, the redox flow batteries, which can withstand frequent charging/discharging operations without aging, were introduced by Sasaki *et al.* [55] for LFC. They have shown faster response compared to other battery ESSs.

On another side, SMES systems have shown superior performance in mitigating the various time scale frequency and tie-line power oscillations in electrical power systems [16, 18]. The small-sized SMES has achieved effectiveness in frequency regulation in two areas of interconnected thermal power systems. Different FLC methods have shown superior performance with SMES systems with different case studies and operational scenarios.

From the control point-of-view, several types of controllers have been applied for VIC using ESSs [56]. The main controllers include the classical PI controllers in addition to the PID control strategies. Additionally, wide applications of the FLC, and the MPC [57, 58], have been introduced with different types of ESSs. Recently, fuzzy type-2 compensators have been applied for LFC in single- and multi-area interconnected MG systems considering the various ESSs. The artificial intelligence-based controllers are also presented in the literature with advanced and/or hybrid control methods to enhance the performance of VIC using ESSs.

The nonlinearities of the GRC and the governor dead band represent an important factor for designing and the optimum tuning of various controllers with ESSs for frequency regulation of electrical power systems [56]. Additionally, further research is targeted for the coordination of the existing various ESSs together and also with the existing digital frequency relays in electrical power systems.

5.5   Control methods of SMES for LFC

Power system operation is maintained stable by regulating system frequency deviations (Δf) with load changes or generation variations [59, 60]. Therefore, the main purpose of the LFC is to keep the power system frequency (50 or 60 Hz) within a specified limit (± 0.5 Hz) by providing/consuming a sufficient amount of energy. In order to optimize the performance of various frequency regulation devices, fast and proper control methods are needed [61]. Additionally, the role of ESSs at high penetration levels of RESs has become essential for the stable and reliable operation of electrical power systems. Additionally, the controller has the function to coordinate among the various connected frequency regulation devices in addition to the protective relaying systems [62]. Several control techniques have been employed in the literature for the frequency regulation function in interconnected electrical power systems: the integer-order controllers, FO controllers, FLC, ANN controllers, MPC methods, SMC methods, etc. [63–65].

On another side, various meta-heuristic optimization techniques are employed for the design of PI controllers, such as genetic algorithms, PSO, marine predator algorithms, hybrid PSO-GA, and Harris Hawks optimization [66].

The optimization methods have the ability to optimize the design process of the PI control methods based on the desired objective function [32, 52]. In which, various optimizers are employed for determining the optimum control parameters. The parameters are determined simultaneously for all the existing controllers in the

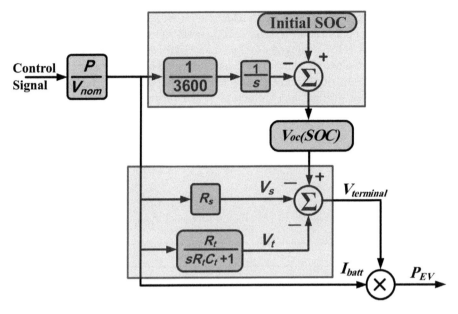

Figure 5.10   The dynamic modeling of EVs for the LFC studies

systems without the need for mathematical modeling and/or complex control design methods [62]. The optimizer has to preserve minimized values of the objective function, which include the various control objectives of frequency deviation in each area in addition to the tie-line power between areas [60].

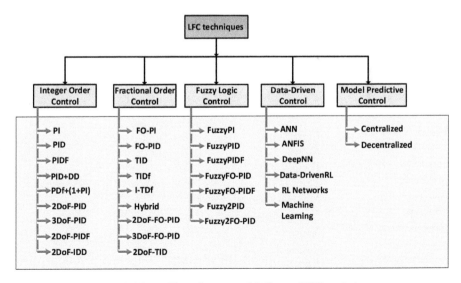

Figure 5.11   Classification of different LFC techniques

The various existing error estimation techniques are vastly utilized in the literature to direct the optimization process. The integral-squared error (ISE), integral-time-squared error (ITSE), integral absolute error (IAE), and integral-time-absolute error (ITAE) and their combinations are extensively used in the literature [67]. They can guarantee the superiority and effectiveness of the controller design by considering the various specifications and limitations of electrical power systems. They can be mathematically modelled as follows:

$$ISE = \int_0^{t_s} \left((\Delta f_a)^2 + (\Delta f_b)^2 + (\Delta P_{tie})^2 \right) dt \tag{5.19}$$

Table 5.1   Main parameters for the studied system ($x \in \{a, b\}$) [104]

Parameter	Symbol	Values	
		Area a	Area b
Area rated capacity	P_{rx}(MW)	1200	1200
Droop constant	R_x(Hz/MW)	2.4	2.4
Frequency bias value	B_x(MW/Hz)	0.4249	0.4249
Minimum limit of valve gate	V_{vlx}(p.u.MW)	−0.5	−0.5
Maximum limit of valve gate	V_{vux}(p.u.MW)	0.5	0.5
Time constant of thermal governor	T_g(s)	0.08	–
Time constant of thermal turbine	T_t(s)	0.3	–
Time constant of hydraulic governor	T_1(s)	–	41.6
Transient droop time constant of hydraulic governor	T_2(s)	–	0.513
Reset time of hydraulic governor	T_R(s)	–	5
Water starting time of hydro turbine	T_w(s)	-	1
Power system inertia constant	H_x(p.u.s)	0.0833	0.0833
Power system damping coefficient	D_x(p.u./Hz)	0.00833	0.00833
Time constant of PV	T_{PV}(s)	–	1.3
Gain of PV	K_{PV}(s)	–	1
Time constant of wind	T_{WT}(s)	1.5	–
Gain of wind	K_{WT}(s)	1	–
Converter time constant of SMES	T_{DCx}(s)	0.03	0.03
Coil of SMES	L_x(H)	0.03	0.03
Control gain of SMES	K_{SMESx}(kV/unit MW)	100	100
Control gain	K_{Idx}(kV/kA)	0.2	0.2
Inductor rated current of SMES	I_{d0x}(kA)	4.5	4.5
Two-area capacity ratio	A_{ab}	−1	
Sync. coefficient of HVDC	$T_{tie,dc}$(s)	0.1732	
Sync. coefficient of HVAC	$T_{tie,ac}$(s)	0.0865	

Table 5.2   Optimal controllers coefficients using MRFO

Controller	Type	Coefficients					
		K_p	K_i	K_d	K_t	λ	μ
LFC	PID_a	1.9647	1.871	1.5242	–	–	–
	PID_b	1.141	0.4278	0.8725	–	–	–
	TID_a	–	1.8869	1.4336	1.9384	1	1
	TID_b	–	0.7102	1.6396	1.509	1	1
	$FOPID_a$	1.8445	1.988	1.4823	–	0.92	0.51
	$FOPID_b$	1.4099	0.0522	0.37968	–	0.94	0.7
	Hybrid (area a)	–	1.9935	1.606	1.9376	0.43	0.28
	Hybrid (area b)	–	0.87343	0.27152	1.5926	0.35	0.18
SMES controller	$SMES_a$	1.6079	0.8966	1.7861	–	1	1
	$SMES_b$	0.8765	0.7709	1.4275	–	1	1

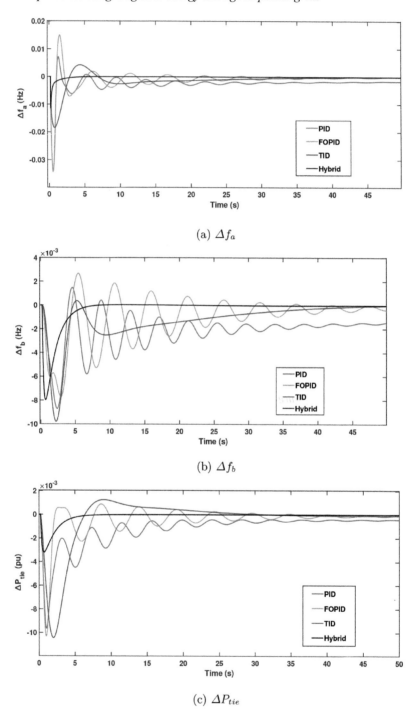

(a) Δf_a

(b) Δf_b

(c) ΔP_{tie}

Figure 5.12   System dynamic response at Scenario 1

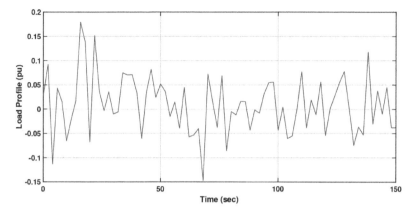

Figure 5.13   Load profile for Scenario 2

$$IAE = \int_{0}^{t_S} (abs(\Delta f_a) + abs(\Delta f_b) + abs(\Delta P_{tie})) \, dt$$

$$(5.20)$$

$$ITSE = \int_{0}^{t_S} ((\Delta f_a)^2 + (\Delta f_b)^2 + (\Delta P_{tie})^2) \, t.dt$$

$$(5.21)$$

$$ITAE = \int_{0}^{t_S} (abs(\Delta f_a) + abs(\Delta f_b) + abs(\Delta P_{tie})) \, t.dt$$

$$(5.22)$$

whereas the optimization process has to limit the determined optimum controller parameters (*PARAM-opt*) based on their corresponding minimum vector of controller parameters (*PARAM-min*) and maximum parameter vector (*PARAM-max*). The MPA determines the optimum parameters within the predefined upper and lower limits of each parameter [67]. The selected limits for the proposed optimization process are as follows:

$$PARAM_{min} \leq PARAM_{opt} \leq PARAM_{max}$$

$$(5.23)$$

5.5.1 Integer-order controllers

The various conventional integer-order control methods have been widely employed for frequency regulation systems. The various combinations and arrangements of the proportional (P), derivative (D), and integral (I) branches have been structured [68, 69]. The classical PI control scheme has been widely utilized for LFC with different power system structures in Reference 70. It has shown vast suitability to control the various single- and multi-area power systems. The PI controller transfer function is represented as follows:

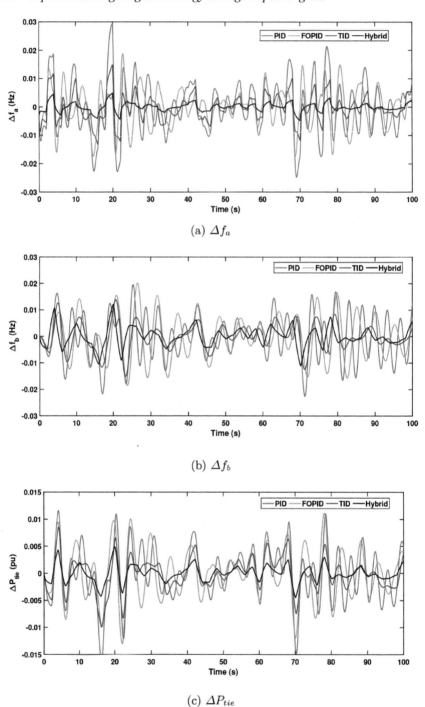

(a) Δf_a

(b) Δf_b

(c) ΔP_{tie}

Figure 5.14   System dynamic response at Scenario 2

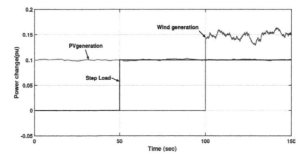

Figure 5.15   Generation and load profiles for Scenario 3

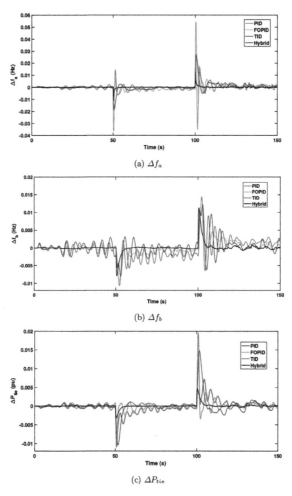

(a) Δf_a

(b) Δf_b

(c) ΔP_{tie}

Figure 5.16   System dynamic response at Scenario 3

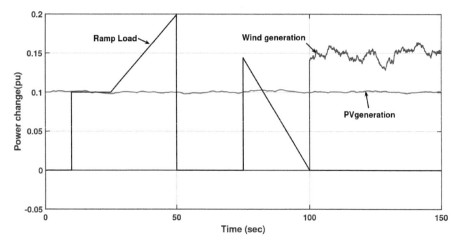

Figure 5.17   *Generation and load profiles for Scenario 4*

$$C(s) = \frac{Y(s)}{R(s)} = K_p + \frac{K_i}{s} \qquad\qquad (5.24)$$

where K_i and K_p represent the PI gains of integral and proportional terms, respectively. The PI control represents a simple solution; however, it fails in the total mitigation of various power system fluctuations. Additionally, it has instability problems when ignoring the time delays of the existing communication systems.

On another side, the PID control method has been widely applied in frequency regulation of electrical power systems. It has shown improved performance, regarding the overcome of frequency instability problem. The various optimization methods have been also applied for designing the PID control parameters. In the study by Sharma *et al.* [71], the various parameters of the PID control method were designed using the stability boundary locus method. The transfer function for the PID control method can be expressed as follows:

$$C(s) = \frac{Y(s)}{R(s)} = K_p + \frac{K_i}{s} + K_d s \qquad\qquad (5.25)$$

where K_d represents the derivative gain of PID controller.

5.5.2 FO controllers

The FO control theory has found wide utilization in the LFC of electrical power systems. The FO integral term, the FO derivative term, and the FO tilt (T) term have been used in various control structures and combinations. In References 72–75, the FO control methods have been introduced and compared in LFC applications. The integer-order PID, FO–PID, two degree of freedom (2DoF)-PID, 2DoF–FO–PID, 3DoF–PID, 3DoF–FO–PID, and cascaded Fuzzy-FO–IDF are included in the comparison.

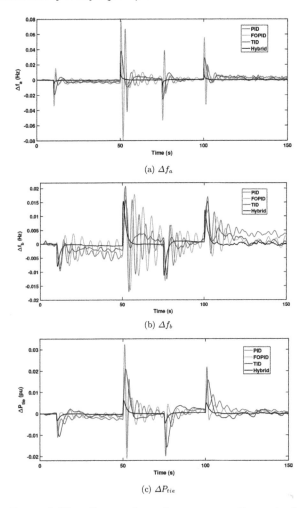

(a) Δf_a

(b) Δf_b

(c) ΔP_{tie}

Figure 5.18   System dynamic response at Scenario 4

On another side, controlling EV systems using the tilt integral derivative (TID) control method has been presented by Oshnoei *et al.* [76], wherein the artificial bee colony optimization scheme is used for determining the controller parameters. Moreover, an optimized design of a modified FO control method has been introduced by Ahmed *et al.* [77] for two-area interconnected electrical power systems. Another application with the PSO optimizer has been proposed by Magdy *et al.* [78] for VIC applications. The pathfinder algorithm has been utilized for optimizing the TID control method for LFC applications by Priyadarshani *et al.* [79]. Moreover, the movable damped wave algorithm has been applied for designing the FO-PID control method by Fathy and Alharbi [80] in multi-area MG. In the study by Ayas and Sahin [81], FO-PID control method has been presented with the sine-cosine optimization algorithm. Although the FO-based control methods cannot damp out wide ranges of frequency and tie-line power fluctuations as they use a single degree of freedom (1DoF).

The FO-PI controller has been presented in the literature for single- and multi-area power systems. It combines the classical P term from integer-order control with the FO integral term. It can be represented as follows:

$$C(s) = \frac{Y(s)}{R(s)} = K_p + \frac{K_i}{s^\lambda} \tag{5.26}$$

where λ denotes the FO of the integral term. It is usually tuned in the range between [0, 1]. On another side, the FO-PID control method has been presented in the literature by combining both merits of the integer-order PID controller with FO calculus. The transfer function of FOPID control method can be expressed as follows:

$$C(s) = \frac{Y(s)}{R(s)} = K_p + \frac{K_i}{s^\lambda} + K_d s^\mu \tag{5.27}$$

where μ denotes the FO of the derivative term. It is usually tuned in the range between [0, 1]. The application of FO-PID control through using the FO parts (*lambda* and μ) results in enhanced dynamic response performance of the LFC system compared to integer-order PID control methods [59].

Furthermore, the TID control method has found wide applications in frequency regulation in electrical power systems. They represent an extended application for the FO control methods in LFC. The mathematical transfer function of TID control method can be represented as follows:

$$C(s) = \frac{Y(s)}{R(s)} = K_t s^{-(\frac{1}{n})} + \frac{K_i}{s} + K_d s \tag{5.28}$$

where K_t represents the tilt term gain. Whereas, n denotes a non-zero real number. It is normally selected in the range between a minimum value of 2.0 and a maximum value of 5.0 in LFC applications [69, 82]. In TID control methods, the various existing gain terms and FO components are optimized to provide a more stable and robust operation of overall electrical power system.

Additionally, modified FO control methods have been presented in the literature. The hybrid FO control method has been introduced by Mohamed *et al.* [59]. The hybrid controller combines both characteristics of TID and FO-PID control methods. It has proven a superior performance and dynamics over the classical PID, TID, and FO-PID control methods. The mathematical transfer function for the hybrid FO control method can be expressed as follows:

$$C(s) = \frac{Y(s)}{R(s)} = K_t s^{-(\frac{1}{n})} + \frac{K_i}{s^\lambda} + K_d s^\mu \tag{5.29}$$

The above-mentioned control methods employ a 1DoF control method for LFC applications. In which, a single feedback is used based on the ACE signal of each area. However, using 2DoF control methods can achieve faster dynamic response and proper mitigation of the frequency and tie-line power fluctuations in electrical power systems. In 2DoF control methods, both the ACE signal and the frequency deviation signal are utilized as feedback signals for the electrical power system. In the study by Ahmed *et al.* [67], a 2DoF FO control method has been proposed for two-area electrical power systems. It uses the hybrid FO control method for the ACE control loop and the TID with filter for the frequency control loop. Thence, the 2DoF control method is capable of damping the high-frequency fluctuations using the frequency control loop and damping the low-frequency fluctuations using the ACE control loop. The mathematical transfer function representation for the 2DoF control method can be represented as follows:

$$\begin{aligned} Y(s) &= (K_{t1} \, s^{-(\frac{1}{n_1})} + \frac{K_{i1}}{s} + K_{d1} \, \frac{N_c s}{s+N_c})\Delta f_x \\ &+ (K_{t2} \, s^{-(\frac{1}{n_2})} + \frac{K_{i2}}{s^{\lambda_1}} + K_{d2} \, s^{\mu_1})ACE_x \end{aligned} \tag{5.30}$$

where N_c denotes the coefficient of derivative filter in the frequency control loop, and ($x \in \{a, b, ...\}$ for interconnected areas).

5.5.3 Fuzzy logic controllers

With the increased power system complexity and uncertainty particularly during load changes and power generation variations, fuzzy logic controllers are considered the most suitable controllers for regulating power system frequency and preserving system stability. Fuzzy logic proportional-integral (FLPI) [83], fuzzy logic PID [84], and neuro-fuzzy [85] controllers are commonly used in multi-area power systems. Fuzzy logic controllers significantly represent better performance compared to the classical fixed gains PI or PID controllers [86]. In addition, combinations of FLC with FO controllers have been demonstrated in LFC applications. The performance of PIDN FO integral-based fuzzy logic (FL-PIDNFOI), fuzzy logic FO integral derivative (FL-FOID), and fuzzy logic FO integral derivative with filter has been evaluated by Arya [72, 87] for maintaining system stability and regulating tie-line power in interconnected power system. Moreover, the integration of FOPI and FOPD using fuzzy logic has been developed on LFC in multi-area power systems [88]. For better performance, many different FO architectures based on fuzzy logic controllers such as fuzzy logic-based FOPD with integral (FL-FOPD+ I) [89], type-2 FO fuzzy PD with fuzzy PI (T2FO-FLPD-FLPI) [90], and fuzzy logic-based FO proportional-integral with FO proportional-integral-derivative (FL-FOPI-FOPID) [91] have been demonstrated in the literature. Recently, different metaheuristic optimization techniques have been applied for optimally designing the scaling factors and the input-output membership function of the fuzzy controller [92–94].

5.5.4 Data-driven LFC methods

Recently, the data-driven LFC methods have found increased applications in electrical power systems [95]. They have shown promising control solutions for the cooperative control of frequency and tie-line power of multi-area interconnected power systems. They aim to the direct optimization of various control criteria using the measured response of power systems [96]. By selecting the proper training protocol in addition to the reward function, the multi-agent reinforcement learning (RL) data-driven LFC methods can achieve cooperative control behavior between areas without the need for full communication among the existing controllers. The classical RL-based control methods by Lillicrap *et al.* [97] achieve the control action using low-dimensional action domains. However, the controller performance is inherently limited by the control action discretization. Whereas, the deep RL-based multi-agent generation control has been proposed for enhancing the control system coordination, wherein the discretization of optimum joint actions is made. Additional continuous actions domain-based deep RL modeling has been proposed by Lowe *et al.* [98] for improving the single-area LFC performance. Another data-driven-based cooperative LFC method has been proposed by Yan and Xu [99] using the multi-agent deep RL method for multi-area interconnected electrical power systems. The presented controller can effectively minimize the frequency fluctuations in each area in addition to mitigating the tie-line power deviations. The presented controller can adjust cooperatively the generated power in each area based on the control action. The presented method uses centralized offline learning to reach the global objective, whereas applying the online control is made individually in each area.

5.5.5   Model predictive controllers

Recently, MPC as a modern controller has been applied in the LFC issues. MPC proves its capability to preserve power system stability during load changes and power generation variation [100]. MPC uses the explicit power system model to estimate the future control response. For proper enhancement of power system stability, MPC has been optimized using a sooty terns optimizer by Elsisi *et al.* and Ali *et al.* [101, 102] and multiverse optimization by Ali *et al.* [103].

The classification of the various addressed LFC techniques in the literature is shown in Figure 5.11.

5.6   Case study

The complete system of the case study under testing has been implemented with the MATLAB Simulink software package. A case study has been developed to explore the performance of LFC with different controller structures. The developed case study consists of two areas: Area 1 includes thermal power plant, wind generator, SMES, and local loads, whereas Area 2 contains hydraulic power plant, PV generators, SMES, and local load as shown in Figure 5.3. Different control structures have been selected for LFC, such as PID, FOPID, TID, and hybrid controller. All the controller parameters have been optimized using MRFO algorithm to minimize the objective function (5.19). Different unit ratings and parameters are presented in Table 5.1. The optimized controllers parameters are shown in Table 5.2. Four scenarios have been developed as follows:

5.6.1   Scenario 1

Figure 5.12 shows the performance of the interconnected power system starting with a 10% step load change in area (a) and area (b). The frequency deviation in area (a) and area (b) is shown in Figure 5.12a and b, respectively, while Figure 5.12c shows the tie-line power deviation between the two areas. It is seen from the figures that the PID controller failed to retain system frequency at the original value. TID has the ability to recover system frequency with a long settling time of about 40 seconds. FOPID has the capability to recover system frequency with over-damped oscillations and long settling time (45 seconds). However, the hybrid controller maintains system frequency with decreased overshoot and settling time compared to the other controllers.

5.6.2   Scenario 2

This scenario presents the LFC performance with a load variation in area (a) as shown the load profile in Figure 5.13. Figure 5.14a, b and c, shows the frequency deviations in area (a), area (b), and tie-line power, respectively. It is seen that the hybrid controller outperforms PID, FOPID, and TID. The PID, FOPID, and TID controllers suffer from frequency oscillations around the steady state value and sometimes higher overshoots appear during load changes.

5.6.3   Scenario 3

This scenario introduces the LFC performance with different variations in load and generations as well. Figure 5.15 presents a 10% variation in PV generation at the start, 10% step load change at 50 seconds, and 15% wind profile changes at 100 seconds. Figure 5.16a, b and c, presents the frequency deviations in area (a), area (b), and tie-line power, respectively. It is seen that the hybrid controller has the ability to keep system frequency at the original values compared to the other controllers. PID, FOPID, and TID suffer from large oscillation around the steady state values of the frequency and the tie-line power as well.

5.6.4   Scenario 4

This scenario investigates the performance of LFC with a more complex scenario as shown in Figure 5.17. Figure 5.18a, b and c presents the frequency deviations in area (a), area (b), and tie-line power, respectively. Almost, the hybrid control gives the best performance compared to the other controllers. The hybrid controller has the capability to damp the frequency oscillations originated from load or generation variations. The other controllers suffer from several oscillations and take a long settling time to overcome such variations.

5.7   Conclusion

This chapter has introduced one of the main applications of SMES in frequency regulation of interconnected MG power systems. The fast response of

SMES devices has made them suitable for fast responding to various genera-tions and loading transients. Moreover, this chapter has introduced the main operation and modeling of renewable energy-based interconnected MG sys-tems with SMES devices. The applications of SMES devices in frequency regulation still need more future research, regarding precise modeling and advanced control techniques for optimizing the operation of SMES. Also, fur-ther research is needed for the coordination of hybrid ESSs with SMES and the other existing controllers and protective relays in order to increase the power system reliability of the interconnected MG systems.

References

[1] Holjevac N., Baškarad T., Đaković J., Krpan M., Zidar M., Kuzle I. 'Challenges of high renewable energy sources integration in power sys-tems—the case of croatia'. *Energies*. 2021, vol. 14(4), p. 1047. Available from https://www.mdpi.com/1996-1073/14/4/1047

[2] Obaid Z.A., Ali Mejeed R., Al-Mashhadani A. 'Investigating the impact of using modern power system stabilizers on frequency stability in large dynamic multi-machine power system'. Presented at 55th International Universities Power Engineering Conference (UPEC) [online]; Torino, Italy, 1–4 September 2020. Available from https://ieeexplore.ieee.org/xpl/mos-tRecentIssue.jsp?punumber=9207782

[3] Nguyen H.T., Yang G., Nielsen A.H., Jensen P.H. 'Frequency stability en-hancement for low inertia systems using synthetic inertia of wind power'. Presented at IEEE Power & Energy Society General Meeting (PESGM); Chicago, IL, 16–20 July 2017.

[4] Said S.M., Aly M., Hartmann B., Alharbi A.G., Ahmed E.M. 'SMES-based fuzzy logic approach for enhancing the reliability of microgrids equipped with pv generators'. *IEEE Access: Practical Innovations, Open Solutions*. 2019, vol. 7, pp. 92059–69.

[5] Aly M., Ahmed E.M., Shoyama M. 'Thermal and reliability assessment for wind energy systems with DSTATCOM functionality in resilient mi-crogrids'. *IEEE Transactions on Sustainable Energy*. 2017, vol. 8(3), pp. 953–65.

[6] Rakhshani E., Remon D., Mir Cantarellas A., Rodriguez P. 'Analysis of derivative control based virtual inertia in multi-area high-voltage direct current interconnected power systems'. *IET Generation, Transmission & Distribution*. 2016, vol. 10(6), pp. 1458–69.

[7] Zhao J., Lyu X., Fu Y., Hu X., Li F. 'Coordinated microgrid frequency regu-lation based on DFIG variable coefficient using virtual inertia and primary frequency control'. *IEEE Transactions on Energy Conversion*. 2015, vol. 31(3), pp. 833–45.

[8] Zhao H., Yang Q., Zeng H. 'Multi-loop virtual synchronous generator con-trol of inverter-based DGS under microgrid dynamics'. *IET Generation, Transmission & Distribution*. 2017, vol. 11(3), pp. 795–803.

[9] D'Arco S., Suul J.A., Fosso O.B. ' Small-signal modeling and parametric sensitivity of a virtual synchronous machine in islanded operation '. *International Journal of Electrical Power & Energy Systems*. 2015, vol. 72, pp. 3–15.

[10] Tamrakar U., Shrestha D., Maharjan M., Bhattarai B., Hansen T., Tonkoski R. 'Virtual inertia: current trends and future directions'. *Applied Sciences*. 2015, vol. 7(7), p. 654.

[11] Kerdphol T., Watanabe M., Mitani Y., Phunpeng V. 'Applying virtual inertia control topology to SMES system for frequency stability improvement of low-inertia microgrids driven by high renewables'. *Energies*. 2019, vol. 12(20), p. 3902.

[12] Liu J., Miura Y., Ise T. 'Comparison of dynamic characteristics between virtual synchronous generator and droop control in inverter-based distributed generators'. *IEEE Transactions on Power Electronics*. 2016, vol. 31(5), pp. 3600–11.

[13] Van de Vyver J., De Kooning J.D.M., Meersman B., Vandevelde L., Vandoorn T.L. 'Droop control as an alternative inertial response strategy for the synthetic inertia on wind turbines'. *IEEE Transactions on Power Systems*. 2016, vol. 31(2), pp. 1129–38.

[14] Chen D., Xu Y., Huang A.Q. 'Integration of DC microgrids as virtual synchronous machines into the AC grid'. *IEEE Transactions on Industrial Electronics*. 2017, vol. 64(9), pp. 7455–66.

[15] Kerdphol T., Rahman F.S., Mitani Y., Watanabe M., Kufeoglu S. 'Robust virtual inertia control of an islanded microgrid considering high penetration of renewable energy'. *IEEE Access*. 2018, vol. 6, pp. 625–36.

[16] Said S.M., Aly M., Hartmann B., Mohamed E.A. 'Coordinated fuzzy logic-based virtual inertia controller and frequency relay scheme for reliable operation of low-inertia power system'. *IET Renewable Power Generation*. 2021, vol. 15(6), pp. 1286–300.

[17] Srinivasarathnam C., Yammani C., Maheswarapu S. 'Load frequency control of multi-microgrid system considering renewable energy sources using grey wolf optimization'. *Smart Science*. 2019, vol. 7(3), pp. 198–217.

[18] Annamraju A., Nandiraju S. 'Robust frequency control in an autonomous microgrid: a two-stage adaptive fuzzy approach'. *Electric Power Components and Systems*. 2018, vol. 46(1), pp. 83–94.

[19] Said S.M., Mohamed E.A., Hartmann B., Mitani Y. 'Coordination strategy for digital frequency relays and energy storage in a low-inertia microgrid'. *Journal of Power Technologies*. 2020, vol. 99(4), pp. 254–63.

[20] Flourentzou N., Agelidis V.G., Demetriades G.D. 'VSC-based HVCD power transmission systems: an overview'. *IEEE Transactions on Power Electronics*. 2021, vol. 24(3), pp. 592–602.

[21] Patra S.P. 'Converting equipment in HVCS transmission'. *IEE-IERE Proceedings - India*. 1972, vol. 10(1), p. 2.

[22] Al-Haiki Z.E., Shaikh-Nasser A.N. 'Power transmission to distant offshore facilities'. *IEEE Transactions on Industry Applications*. 2011, vol. 47(3), pp. 1180–83.

[23]   Ibraheem K.P. 'Current status of the indian power system and dynamic performance enhancement of hydro power systems with asynchronous tie lines'. *Electric Power Components and Systems*. 1972, vol. 31(7), pp. 605–26.

[24]   Ngamroo I. 'A stabilization of frequency oscillation in a parallel AC-DC interconnected power system via an HYDC link'. *ScienceAsia*. 2002, vol. 28(2), p. 173.

[25]   Thottungal R., Anbalagan P., Mohanaprakash T., Sureshkumar A., Prabhu G.V. 'Frequency stabilisation in multi area system using HVDC link'. *2006 IEEE International Conference on Industrial Technology*; Mumbai, India, 2006. pp. 590–95.

[26]   El-emary A.A., El-shibina M.A. 'Application of static VAR compensation for load frequency control'. *Electric Machines & Power Systems*. 1997, vol. 25(9), pp. 1009–22.

[27]   Shankar R., Chatterjee K. 'Facts based controller for interconnected hydro-thermal power system'. *International Journal of Engineering Science and Technology*. 2012, vol. 4(4), pp. 1776–86.

[28]   Chaudhuri B., Pal B.C., Zolotas A.C., Jaimoukha I.M., Green T.C. 'Mixed-sensitivity approach to h//sub∞/ control of power system oscillations employing multiple facts devices'. *IEEE Transactions on Power Systems*. 1997, vol. 18(3), pp. 1149–56.

[29]   Bhatt P., Ghoshal S.P., Roy R. 'Load frequency stabilization by coordinated control of thyristor controlled phase shifters and superconducting magnetic energy storage for three types of interconnected two-area power systems'. *International Journal of Electrical Power & Energy Systems*. 1997, vol. 32(10), pp. 1111–24.

[30]   Ngamroo I., Tippayachai J., Dechanupaprittha S. 'Robust decentralised frequency stabilisers design of static synchronous series compensators by taking system uncertainties into consideration'. *International Journal of Electrical Power & Energy Systems*. 2006, vol. 28(8), pp. 513–24.

[31]   Bhongade S., Eappen G., Gupta R.O. 'Coordination control scheme by SSSC and TCPS with redox flow battery for optimized automatic generation control'. *2013 International Conference on Renewable Energy and Sustainable Energy (ICRESE)*; Coimbatore, 2013. pp. 145–50.

[32]   Fernández-Guillamón A., Gómez-Lázaro E., Muljadi E., Molina-García Á. ' power systems with high renewable energy sources: a review of inertia and frequency control strategies over time '. *Renewable and Sustainable Energy Reviews*. 2019, vol. 115, p. 109369.

[33]   Moutis P., Vassilakis A., Sampani A., Hatziargyriou N.D. 'DC switch driven active power output control of photovoltaic inverters for the provision of frequency regulation'. *IEEE Transactions on Sustainable Energy*. 2014, vol. 6(4), pp. 1485–93.

[34]   Zarina P.P., Mishra S., Sekhar P.C. 'Exploring frequency control capability of a PV system in a hybrid PV-rotating machine-without storage system'. *International Journal of Electrical Power & Energy Systems*. 2014, vol. 60, pp. 258–67.

[35] Rahmann C., Castillo A. 'Fast frequency response capability of photovoltaic power plants: the necessity of new grid requirements and definitions'. *Energies*. 2014, vol. 7(10), pp. 6306–22.

[36] Alatrash H., Mensah A., Mark E., Haddad G., Enslin J. 'Generator emulation controls for photovoltaic inverters'. *IEEE Transactions on Smart Grid*. 2012, vol. 3(2), pp. 996–1011.

[37] Mishra S., Zarina P.P., Sekhar P.C. 'A novel controller for frequency regulation in A hybrid system with high pv penetration'. *2013 IEEE Power & Energy Society General Meeting*; Vancouver, BC, Canada, 2013. pp. 1–5.

[38] Zarina P.P., Mishra S., Sekhar P.C. 'Deriving inertial response from a non-inertial PV system for frequency regulation'. *2012 IEEE International Conference on Power Electronics, Drives and Energy Systems (PEDES)*; Bengaluru, India, 2012. pp. 1–5.

[39] Ma H.T., Chowdhury B.H. 'Working towards frequency regulation with wind plants: combined control approaches'. *IET Renewable Power Generation*. 2012, vol. 4(4), p. 308.

[40] Moutis P., Papathanassiou S.A., Hatziargyriou N.D. 'Improved load-frequency control contribution of variable speed variable pitch wind generators'. *Renewable Energy*. 2012, vol. 48, pp. 514–23.

[41] Vidyanandan K.V., Senroy N. 'Primary frequency regulation by deloaded wind turbines using variable droop'. *IEEE Transactions on Power Systems*. 2007, vol. 28(2), pp. 837–46.

[42] Ramtharan G., Jenkins N., Ekanayake J.B. 'Frequency support from doubly fed induction generator wind turbines'. *IET Renewable Power Generation*. 2007, vol. 1(1), p. 3.

[43] Huang L., Xin H., Zhang L., Wang Z., Wu K., Wang H. 'Synchronization and frequency regulation of DFIG-based wind turbine generators with synchronized control'. *IEEE Transactions on Energy Conversion*. 2007, vol. 32(3), pp. 1251–62.

[44] Dreidy M., Mokhlis H., Mekhilef S. 'Inertia response and frequency control techniques for renewable energy sources: a review'. *Renewable and Sustainable Energy Reviews*. 2017, vol. 69, pp. 144–55.

[45] Salama H.S., Said S.M., Aly M., Vokony I., Hartmann B. 'Studying impacts of electric vehicle functionalities in wind energy-powered utility grids with energy storage device'. *IEEE Access: Practical Innovations, Open Solutions*. 2007, vol. 9, pp. 45754–69.

[46] Khan M., Sun H., Xiang Y., Shi D. 'Electric vehicles participation in load frequency control based on mixed h2/h∞'. *International Journal of Electrical Power & Energy Systems*. 2014, vol. 125, p. 106420.

[47] Bernard M.Z., Mohamed T.H., Qudaih Y.S., Mitani Y. 'Decentralized load frequency control in an interconnected power system using coefficient diagram method'. *International Journal of Electrical Power & Energy Systems*. 2014, vol. 63, pp. 165–72.

[48] Mohamed E.A., Ahmed E.M., Elmelegi A., Aly M., Elbaksawi O., Mohamed A.-A.A. 'An optimized hybrid fractional order controller for

frequency regulation in multi-area power systems'. *IEEE Access: Practical Innovations, Open Solutions.* 2014, vol. 8, pp. 213899–915.

[49] Ota Y., Taniguchi H., Nakajima T., Liyanage K.M., Baba J., Yokoyama A. 'Autonomous distributed v2g (vehicle-to-grid) satisfying scheduled charging'. *IEEE Transactions on Smart Grid.* 2014, vol. 3(1), pp. 559–64.

[50] Falahati S., Taher S.A., Shahidehpour M. 'Grid secondary frequency control by optimized fuzzy control of electric vehicles'. *IEEE Transactions on Smart Grid.* 2014, vol. 9(6), pp. 5613–21.

[51] Luo X., Xia S., Chan K.W. 'A decentralized charging control strategy for plug-in electric vehicles to mitigate wind farm intermittency and enhance frequency regulation'. *Journal of Power Sources.* 2004, vol. 248, pp. 604–14.

[52] Shankar R., Pradhan S.R., Chatterjee K., Mandal R. 'A comprehensive state of the art literature survey on LFC mechanism for power system'. *Renewable and Sustainable Energy Reviews.* 2017, vol. 76, pp. 1185–207.

[53] Aditya S.K., Das D. 'Battery energy storage for load frequency control of an interconnected power system'. *Electric Power Systems Research.* 2004, vol. 58(3), pp. 179–85.

[54] Kunisch H.-J., Kramer K.G., Dominik H. 'Battery energy storage another option for load-frequency-control and instantaneous reserve'. *IEEE Transactions on Energy Conversion.* 2004, vol. EC-1 (3), pp. 41–46.

[55] Sasaki T., Kadoya T., Enomoto K. 'Study on load frequency control using redox flow batteries'. *IEEE Transactions on Power Systems.* 2004, vol. 19(1), pp. 660–67.

[56] Dreidy M., Mokhlis H., Mekhilef S. 'Inertia response and frequency control techniques for renewable energy sources: a review'. *Renewable and Sustainable Energy Reviews.* 2017, vol. 69, pp. 144–55.

[57] M. Said S., Ali A., Hartmann B. 'Tie-line power flow control method for grid-connected microgrids with SMES based on optimization and fuzzy logic'. *Journal of Modern Power Systems and Clean Energy.* 2017, vol. 8(5), pp. 941–50.

[58] Kerdphol T., Rahman F., Mitani Y., Hongesombut K., Küfeoğlu S. 'Virtual inertia control-based model predictive control for microgrid frequency stabilization considering high renewable energy integration'. *Sustainability.* 2017, vol. 9(5), p. 773.

[59] Mohamed E.A., Ahmed E.M., Elmelegi A., Aly M., Elbaksawi O., Mohamed A.-A.A. 'An optimized hybrid fractional order controller for frequency regulation in multi-area power systems'. *IEEE Access: Practical Innovations, Open Solutions.* 2017, vol. 8, pp. 213899–915.

[60] Pappachen A., Peer Fathima A. 'Critical research areas on load frequency control issues in a deregulated power system: a state-of-the-art-of-review'. *Renewable and Sustainable Energy Reviews.* 2017, vol. 72, pp. 163–77.

[61] Akram U., Nadarajah M., Shah R., Milano F. 'A review on rapid responsive energy storage technologies for frequency regulation in modern power systems'. *Renewable and Sustainable Energy Reviews.* 2020, vol. 120, p. 109626.

[62] Elmelegi A., Mohamed E.A., Aly M., Ahmed E.M., Mohamed A.-A.A., Elbaksawi O. 'Optimized tilt fractional order cooperative controllers for preserving frequency stability in renewable energy-based power systems'. *IEEE Access: Practical Innovations, Open Solutions*. 2020, vol. 9, pp. 8261–77.

[63] Dreidy M., Mokhlis H., Mekhilef S. 'Inertia response and frequency control techniques for renewable energy sources: a review'. *Renewable and Sustainable Energy Reviews*. 2020, vol. 69, pp. 144–55.

[64] Pandey S.K., Mohanty S.R., Kishor N. 'A literature survey on load–frequency control for conventional and distribution generation power systems'. *Renewable and Sustainable Energy Reviews*. 2020, vol. 25, pp. 318–34.

[65] Shankar R., Pradhan S.R., Chatterjee K., Mandal R. 'A comprehensive state of the art literature survey on LFC mechanism for power system'. *Renewable and Sustainable Energy Reviews*. 2020, vol. 76, pp. 1185–207.

[66] Yousri D., Babu T.S., Fathy A. 'Recent methodology based harris hawks optimizer for designing load frequency control incorporated in multi-interconnected renewable energy plants'. *Sustainable Energy, Grids and Networks*. 2020, vol. 22, p. 100352.

[67] Ahmed E.M., Elmelegi A., Shawky A., Aly M., Alhosaini W., Mohamed E.A. 'Frequency regulation of electric vehicle-penetrated power system using mpa-tuned new combined fractional order controllers'. *IEEE Access: Practical Innovations, Open Solutions*. 2020, vol. 9, pp. 107548–65.

[68] Nanda J., Sreedhar M., Dasgupta A. 'A new technique in hydro thermal interconnected automatic generation control system by using minority charge carrier inspired algorithm'. *International Journal of Electrical Power & Energy Systems*. 2020, vol. 68, pp. 259–68.

[69] Singh K., Amir M., Ahmad F., Khan M.A. 'An integral tilt derivative control strategy for frequency control in multi microgrid system'. *IEEE Systems Journal*. 2020, vol. 15(1), pp. 1477–88.

[70] Khalil A., Rajab Z., Alfergani A., Mohamed O. 'The impact of the time delay on the load frequency control system in microgrid with plug-in-electric vehicles'. *Sustainable Cities and Society*. 2020, vol. 35, pp. 365–77.

[71] Sharma J., Hote Y.V., Prasad R. 'PID controller design for interval load frequency control system with communication time delay'. *Control Engineering Practice*. 2019, vol. 89, pp. 154–68.

[72] Arya Y. 'Effect of electric vehicles on load frequency control in interconnected thermal and hydrothermal power systems utilising CF-FOIDF controller'. *IET Generation, Transmission & Distribution*. 2020, vol. 14(14), pp. 2666–75.

[73] Zamani A., Barakati S.M., Yousofi-Darmian S. 'Design of a fractional order PID controller using GBMO algorithm for load-frequency control with governor saturation consideration'. *ISA Transactions*. 2016, vol. 64, pp. 56–66.

[74] Debbarma S., Dutta A. 'Utilizing electric vehicles for LFC in restructured power systems using fractional order controller'. *IEEE Transactions on Smart Grid*. 2019, vol. 8(6), pp. 2554–64.

[75]   Nayak J.R., Shaw B., Sahu B.K. 'Implementation of hybrid SSA–SA based three-degree-of-freedom fractional-order PID controller for AGC of a two-area power system integrated with small hydro plants'. *IET Generation, Transmission & Distribution*. 2020, vol. 14(13), pp. 2430–40.

[76]   Oshnoei A., Khezri R., Muyeen S.M., Oshnoei S., Blaabjerg F. 'Automatic generation control incorporating electric vehicles'. *Electric Power Components and Systems*. 2019, vol. 47(8), pp. 720–32.

[77]   Ahmed E.M., Mohamed E.A., Elmelegi A., Aly M., Elbaksawi O. 'Optimum modified fractional order controller for future electric vehicles and renewable energy-based interconnected power systems'. *IEEE Access: Practical Innovations, Open Solutions*. 2019, vol. 9, pp. 29993–30010.

[78]   Magdy G., Bakeer A., Nour M., Petlenkov E. 'A new virtual synchronous generator design based on the SMES system for frequency stability of low-inertia power grids'. *Energies*. 2019, vol. 13(21), p. 5641.

[79]   Priyadarshani S., Subhashini K.R., Satapathy J.K. 'Pathfinder algorithm optimized fractional order tilt-integral-derivative (FOTID) controller for automatic generation control of multi-source power system'. *Microsystem Technologies*. 2021, vol. 27(1), pp. 23–35.

[80]   Fathy A., Alharbi A.G. 'Recent approach based movable damped wave algorithm for designing fractional-order PID load frequency control installed in multi-interconnected plants with renewable energy'. *IEEE Access: Practical Innovations, Open Solutions*. 2019, vol. 9, pp. 71072–89.

[81]   Ayas M.S., Sahin E. 'FOPID controller with fractional filter for an automatic voltage regulator'. *Computers & Electrical Engineering*. 2021, vol. 90, p. 106895.

[82]   Kumar Sahu R., Panda S., Biswal A., Chandra Sekhar G.T. 'Design and analysis of tilt integral derivative controller with filter for load frequency control of multi-area interconnected power systems'. *ISA Transactions*. 2016, vol. 61, pp. 251–64.

[83]   Sahoo D.K., Sahu R.K., Sekhar G.T.C., Panda S. 'A novel modified differential evolution algorithm optimized fuzzy proportional integral derivative controller for load frequency control with thyristor controlled series compensator'. *Journal of Electrical Systems and Information Technology*. 2021, vol. 5(3), pp. 944–63.

[84]   Pothiya S., Ngamroo I. 'Optimal fuzzy logic-based pid controller for load–frequency control including superconducting magnetic energy storage units'. *Energy Convers Manag*. 2008, vol. 49(10), pp. 2833–38.

[85]   Rahim S.A., Ahmed S., Nawari M. 'A study of load frequency control for two area power system using two controllers'. *2018 International Conference on Computer, Control, Electrical, and Electronics Engineering (ICCCEEE)*; Khartoum, Khartoum, 2021. pp. 1–6.

[86]   Kocaarslan İ., Çam E. 'Fuzzy logic controller in interconnected electrical power systems for load-frequency control'. *International Journal of Electrical Power & Energy Systems*. 2005, vol. 27(8), pp. 542–49.

[87] Arya Y. 'Automatic generation control of two-area electrical power systems via optimal fuzzy classical controller'. *Journal of the Franklin Institute.* 2018, vol. 355(5), pp. 2662–88.

[88] Vigya M.T., Malik H., Mukherjee V., Alotaibi M.A., Almutairi A. 'Renewable generation based hybrid power system control using fractional order-fuzzy controller'. *Energy Reports.* 2021, vol. 7, pp. 641–53.

[89] Khooban M.-H., Dragicevic T., Blaabjerg F., Delimar M. 'Shipboard microgrids: A novel approach to load frequency control'. *IEEE Transactions on Sustainable Energy.* 2005, vol. 9(2), pp. 843–52.

[90] Gheisarnejad M., Khooban M.-H., Dragicevic T. 'The future 5G network-based secondary load frequency control in shipboard microgrids'. *IEEE Journal of Emerging and Selected Topics in Power Electronics.* 2005, vol. 8(1), pp. 836–44.

[91] Arya Y. 'A novel CFFOPI-FOPID controller for AGC performance enhancement of single and multi-area electric power systems'. *ISA Transactions.* 2020, vol. 100, pp. 126–35.

[92] Jalali N., Razmi H., Doagou-Mojarrad H. 'Optimized fuzzy self-tuning PID controller design based on tribe-de optimization algorithm and rule weight adjustment method for load frequency control of interconnected multi-area power systems'. *Applied Soft Computing.* 2020, vol. 93, 106424.

[93] Sahu B.K., Pati T.K., Nayak J.R., Panda S., Kar S.K. 'A novel hybrid LUS–TLBO optimized fuzzy-PID controller for load frequency control of multi-source power system'. *International Journal of Electrical Power & Energy Systems.* 2020, vol. 74, pp. 58–69.

[94] Sahu R.K., Panda S., Chandra Sekhar G.T. 'A novel hybrid PSO-PS optimized fuzzy PI controller for AGC in multi area interconnected power systems'. *International Journal of Electrical Power & Energy Systems.* 2020, vol. 64, pp. 880–93.

[95] Ding Y., Xu N., Ren L., HaoK. 'Data-driven neuroendocrine ultrashort feedback-based cooperative control system'. *IEEE Transactions on Control Systems Technology.* 2020, vol. 23(3), pp. 1205–12.

[96] Zhang C., Lesser V.R. 'Coordinating multi-agent reinforcement learning with limited communication' in *International Conference on Autonomous Agents and Multi-agent Systems (AMAS)*; 2013. pp. 1101–08.

[97] Lillicrap T.P., Hunt J.J., Pritzel A., *et al.* 'Continuous control with deep reinforcement learning'. *CoRR.* 2016, vol. abs/1509, p. 02971.

[98] Lowe R., Wu Y., Tamar A., Harb J., Abbeel P., Mordatch I. 'Multi-agent actor-critic for mixed cooperative-competitive environments' in *31st International Conference on Neural Information Processing Systems*; 2017. pp. 6382–93.

[99] Yan Z., Xu Y. 'A multi-agent deep reinforcement learning method for cooperative load frequency control of a multi-area power system'. *IEEE Transactions on Power Systems.* 2018, vol. 35(6), pp. 4599–608.

[100] Ismail M.M., Bendary A.F. 'Load frequency control for multi area smart grid based on advanced control techniques'. *Alexandria Engineering Journal*. 2018, vol. 57(4), pp. 4021–32.

[101] Elsisi M., Soliman M., Aboelela M.A.S., Mansour W. 'Model predictive control of plug-in hybrid electric vehicles for frequency regulation in a smart grid'. *IET Generation, Transmission & Distribution*. 2017, vol. 11(16), pp. 3974–83.

[102] Ali H.H., Fathy A., Kassem A.M. ' optimal model predictive control for lfc of multi-interconnected plants comprising renewable energy sources based on recent sooty terns approach '. *Sustainable Energy Technologies and Assessments*. 2020, vol. 42, p. 100844.

[103] Ali H.H., Kassem A.M., Al-Dhaifallah M., Fathy A. ' Multi-verse optimizer for model predictive load frequency control of hybrid multi-interconnected plants comprising renewable energy '. *IEEE Access: Practical Innovations, Open Solutions*. 2020, vol. 8, pp. 114623–42.

[104] Abraham R.J., Das D., Patra A. 'Automatic generation control of an interconnected hydrothermal power system considering superconducting magnetic energy storage'. *International Journal of Electrical Power & Energy Systems*. 2007, vol. 29(8), pp. 571–79.

Chapter 6

Dynamic performance enhancement of power grids by coordinated operation of SMES and other control systems

Mohammad Ashraf Hossain Sadi[1]

6.1 Introduction

Power systems have experienced a dramatic change during the past decades due to increased integration of inverter-based resources, increased interconnections, change in consumer load profiles, etc. These advancements and the concept of microgrid have introduced several challenges to the power system operators like weakly damped swings between synchronous generators, less inertia in the system, increased oscillations, sensitivity to grid disturbances, etc. [1]. The Superconducting Magnetic Energy Storage (SMES) as an auxiliary device can tackle the transient disturbances and the oscillations of the power grid by swiftly exchanging active and reactive power with the Alternating Current (AC) system [2, 3]. However, as the power grid becomes complex and diverse, the single alone operation of the SMES may not provide expected reliability, and flexibility. So, to add more degree of freedom, it is imperative that the SMES operate in conjunction with other auxiliary devices so that the advantages of each individual auxiliary devices can be utilized to achieve the expected stability of the power system. The flexible AC transmission systems (FACTS) devices like static var compensator (SVC), static synchronous compensator (STATCOM), etc. are developed and put into operation to improve the transient stability margin, damping the power system oscillations by controlling the reactive power, etc. [4]. Moreover, the SMES not only has high power density but low energy density that can be compensated by battery energy storage systems (BESSs) [5]. In addition to these auxiliary devices and battery energy systems, the fault current limiter (FCL) can reduce the higher transient current in the power grid and improve the dynamic stability [6]. Due to these advantages of the auxiliary devices, this chapter discusses the combined operation of SMES and other auxiliary devices, BESSs, and FCL in improving the dynamic performance of the power grid. The operation principles of the SMES and different auxiliary devices have been

[1]University of Central Missouri, Warrensburg, Missouri, USA

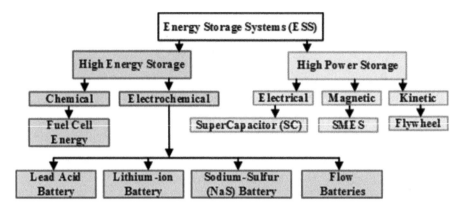

Figure 6.1   Classification of energy storage technologies [7, 8]

discussed with their control systems. Furthermore, different case studies are also carried out to provide better visibility of the results.

6.2   Hybrid energy storage systems (HESS)

Energy storage systems (ESS) pose many advantages in power systems like balancing generation and demand, power quality improvement, frequency and voltage regulation, smoothing the intermittency of the renewable energy sources, etc. [7, 8]. The ESSs can be divided into two different groups based on their power density and energy density capabilities [7, 9] as represented in Figure 6.1.

As can be seen from the classification of the energy storage technologies, power density and energy density are the two main characteristics of the ESSs. However, a single energy storage device cannot fulfill both the power density and energy density requirements and other desired operations due to their limited capabilities and limited responses. Characteristics of different energy storage technologies are presented in Table 6.1.

Energy storage technologies like fuel cell, batteries with lower power density and higher energy density pose power control challenges as the dynamic response of these technologies is slow. Therefore, they can tackle the low-frequency oscillations but not able to minimize the high-frequency oscillations [10, 11]. On the other hand, energy storage technologies like SMES, super capacitors can supply high power demand but unable to support high energy demands. So, high power density ESSs can compensate the swift duration events appropriately whereas high energy density ESSs can compensate the long duration variations.

As none of the available energy storage technologies can meet all the requirements due to their limitations, it is often necessary to enrich the transient, steady, and dynamic performance of the ESSs by combining the advantages of two or more energy storage technologies termed as HESSs. Depending on the requirements and purpose of the hybridization, different energy storage devices can be used as

Table 6.1 Characteristics of different energy storage technologies

Type of storage	Energy density	Power density	Lifetime	Reaction time	Cost	Efficiency
SMES	Low	High	Long	Fast	High	High
Super capacitor (SC)	Low	High	Long	Fast	Medium	Medium
Flywheel	Low	High	Long	Fast	High	High
Fuel cell (FC) energy	High	Low	Long	Slow	Low	Low
Sodium–sulfur (NaS) battery	High	Low	Medium	Fast	Medium	Medium
Lithium-ion battery	High	Low	Medium	Fast	Medium	High

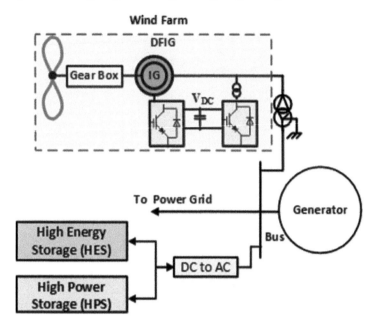

*Figure 6.2   Hybrid energy storage system in the presence of high energy storage
and high power storage*

HESS. Normally, the HESS application consists of high power storage (HPS) and
high energy storage HES devices where one of them meets the long-term energy
demands and the other one meets the transient power demands [12] as represented
in Figure 6.2.

The HESSs in power system improve the system efficiency, reduce the cost,
and prolong the lifetime of the individual ESS. The overall design and selection of
the HESS depends on the choice of storage technology, capacity, power converter
topologies, and their control strategies. Thus, wide range of energy storage device
hybridizations are possible as represented in Figure 6.3.

HESS can be connected directly in the DC (Direct Current) bus of the power
grid. HESS can also be connected to power grid through different interconnection
topologies employing HPS, HES, and power converters. Different connection topologies of the HPS and HES offer better control flexibility, dynamic performance, and
better efficiency. The interconnection topologies are mainly categorized as passive,
semi-active, and active connections [13].

Passive interconnection is the simplest approach where two ESS with same
voltage levels are connected without employing power converters for each of them.
It is the simplest, efficient, and cost-effective topology. The passive interconnection
topology employing HPS and HES is represented in Figure 6.4.

In the semi-active interconnection topology, a bidirectional DC–DC converter
is inserted at the terminal of one of the storages, while the other one is directly connected to the DC bus. Most commonly, the DC–DC converter is inserted after the

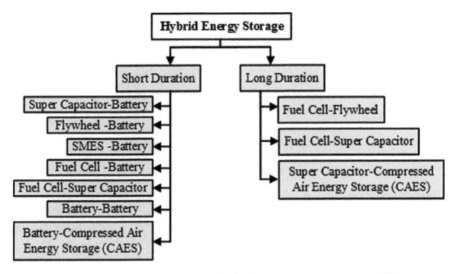

Figure 6.3   Different types of hybrid energy storage systems [9]

HPS so that the peak power requirement of the system is satisfied by the HPS and the remaining demand is met by HES. When the DC–DC converter is connected across the terminal of the HES, then that interconnection is termed as series active topology [7, 9]. Although both these interconnection schemes increase the cost, it comes up with better controllability and dispatch capability. The connection scheme of both the topologies is presented in Figure 6.5.

In active interconnection topology, each ESS employs bidirectional DC–DC converter and connected to DC bus as represented in Figure 6.6. This topology offers highest possible controllability with decouples active controller for both ESS. Thus, this topology can accommodate wide variety of control logics although it comes with the expense of higher converter loss and extra cost for the converters. But the voltage levels of the involved ESSs can be independent of the system voltage and the individual converters have the inherent fault tolerant capability. Each interconnection topologies have its own advantages and disadvantages, but in recent years, the active interconnection topology is getting more application due to its higher capability.

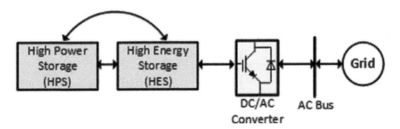

Figure 6.4   Passive interconnection topology of the HESS [9]

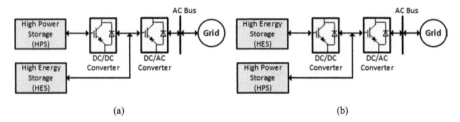

Figure 6.5   (a) Semi-active interconnection topology; (b) series active interconnection topology [7, 8]

Designing and implementing a proper control system for the HESS is the most important issue and the selection of proper control technique relies upon different parameters like power quality, response time, lifetime, structure of hybridization, etc. To achieve the goal of optimized performance of the HESS, an array of control techniques are available, and they are categorized as classical and intelligent control techniques.

6.3   Dynamic performance enhancement of power grids by combination of SMES and battery energy storage

Compared with the traditional fossil fuel-based energy sources, the renewable energy sources cannot provide stable and firm electric power due to their fluctuating nature. ESSs/devices with high energy density and fast response time or high power density are desirable in compensating these fluctuating outputs. Moreover, due to vast advancement and implementation of the microgrids, there happen to be instances when the microgrids have to deal with the high amount of power demand during islanded mode of operation. These situations can also be tackled by the ESS placed in the microgrid level.

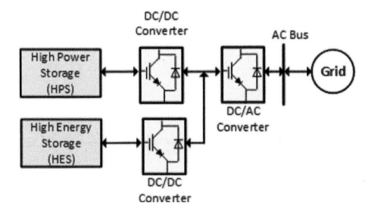

Figure 6.6   Active interconnection topology of the HESS [7, 8]

The HESS combines the functional advantages of two existing ESS and thus provides desired performance and better power and energy performances. Among the available ESSs, the SMES has higher power density, rapid transient response, and good charging-discharging efficiency. Thus, it is desirable for handling short-term instantaneous power demand. On the other hand, the BESS are suitable for application in renewable energy systems and microgrids due to their ability to supply long-term continuous power, higher energy density, better efficiency, robustness, and long-life cycle [14]. Besides, the BESS can take the form of lead–acid battery, lithium-ion battery, sodium–sulfur (NaS) battery, vanadium redox battery, flow batteries, etc. [15]. Since their characteristics are complementary to each other, HESS in the combined form of SMES and BESS will be techno-economically more beneficial for application in power grid, and microgrid [16]. Thus, their combined application will be more effective than a single ESS.

In microgrids, the ESSs are integrated in the system to compensate for the load power demand when the microgrids are disconnected from the main grid and they operate in the islanded mode. However, in many microgrid systems only BESS are used as the ESS, which poses different load demand coordination issues. If the microgrid operates in the islanded mode, then the instantaneous high power load demand cannot be supported by the BESS alone, which is not capable to respond the high transient power demand. In this situation, the battery's lifetime will be attenuated. However, hybridization of the high power energy storage device like the SMES with the BESS can solve this problem. Thus, the HESS combining the operations of both SMES and BESS can deal with the short-term high power demand transients as well as long-term power demand fluctuations of a microgrid in the islanded mode of operation [17]. The SMES–BESS HESSs connection arrangement into the power grid is represented in Figure 6.7.

The initial SMES and BESS are both connected to the DC bus through the DC–DC converter. Their initial objective of the HESS is to maintain the DC bus voltage within a certain range. The bidirectional DC–DC converter is used to control the charging and discharging cycle of the BESS. On the other hand, the DC chopper/DC–DC converter of the SMES is used to support the power exchange between the SMES and the DC bus [18].

The DC chopper of the SMES is designed to control the energy stored in its coil magnet. The DC copper/DC–DC converter comprises of diodes (D1, D2), insulated-gate bipolar transistor (IGBT) switches (S1, S2), and an output capacitor (C) [19, 20] as shown in Figure 6.8 with a simple control system.

For the DC chopper, three operating modes are used that are controlled by the control system: charging mode, discharging mode, and standby mode [21]. Their corresponding current paths are presented in Figure 6.8. The charging mode works when the controller requests SMES to absorb energy from the DC bus. During discharge mode, energy from the SMES is delivered to the DC bus and AC grid. When the SMES does not need to exchange power to and from the DC bus, the DC chopper operates in the standby mode to keep the energy stored in the SMES coil.

The bidirectional DC–DC converter of the battery can track the capacity, has robust performance, and is capable of restraining long-term low-frequency power

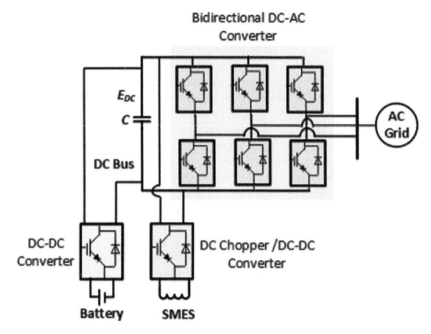

Figure 6.7   SMES–BESS hybrid energy storage systems connection topology [18]

fluctuations in the point of common coupling (PCC). The DC–DC converter of the battery is shown in Figure 6.9.

The BESS will cooperate with the SMES and will act as a backup support for the SMES when its energy becomes insufficient. The DC–DC converter works on the basic boost mode to achieve the desired output voltage. In boost mode, the duty cycle of the IGBT2 is controlled to stabilize the desired output voltage. Furthermore, the DC–DC converter of the battery is used to control the charging and discharging states of the BESS.

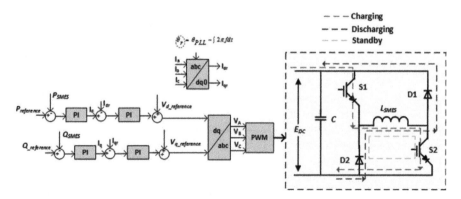

Figure 6.8   SMES DC chopper with its current paths and control system [14]

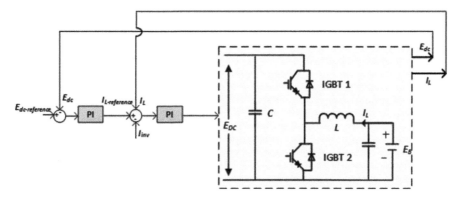

Figure 6.9   DC–DC converter of the BESS

To achieve the proper control of the SMES–BESS hybridization, several control topology can be utilized like: proportional integral (PI) control, filtration method, fuzzy logic control, power grading control, droop control, model predictive control, two stage control method, hysteresis current control, etc.

Besides the SMES–BESS hybridization application in power systems and microgrids, they also have been utilized for several other situations like high fluctuating demands in the electric bus, electric ships, AC systems primary frequency controls, balancing the power demand between the loads and generators, etc.

6.4   Dynamic performance enhancement of power grids by combination of SMES and fuel cell system

Fuel cell is one of the most promising environmentally sound renewable energy sources that have higher efficiency, better scalability. Fuel cell is a static electro-chemical device that converts chemical energy of hydrogen and oxygen directly into electrical energy through the electro-chemical reactions [22]. It can produce electric energy if the chemical energy is supplied to the system and it does not need recharging to produce electricity unlike battery cells. However, there are still some drawbacks of the fuel cell systems like low output voltage, reduced efficiency with ripple current, slow response time, limited overload capacities, and no acceptance to reverse directions currents [23]. These obstacles pose different technical challenges like the requirement of power electronic interfaces to boost up the low voltage, proper utilization of power conditioning systems, etc. Moreover, during grid-tied applications, the fuel cell system cannot respond to sudden changes in active and reactive power demands. But the fuel cell systems are suitable for application in renewable energy systems due to their ability to supply long-term continuous power, and higher energy density [24].

As the SMES has higher power density, rapid transient response, and good charging-discharging efficiency, it can compensate the transient shocks in the power grid caused by sudden change in active and reactive power demands. The combined

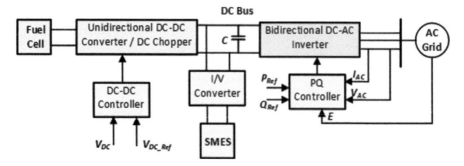

Figure 6.10   *Grid-connected SMES–fuel cell combined operation scheme [22, 25]*

operation of SMES and fuel cell can combine the advantages of both the devices and provide the high power and high energy solutions for reducing transient influence cause by sudden change in active and reactive power load demands in the power grid. In their combined operation, the SMES will compensate the short-term high power demand transients, while the fuel cell system will compensate the long-term power demand fluctuations of the power system. Grid-connected combined operation scheme of the SMES and fuel cell is provided in Figure 6.10.

The unidirectional DC–DC converter or the DC chopper is used to boost the low DC voltage of the fuel cells into the higher voltage and provide electrical isolation of the low and high voltage circuits. The bidirectional DC–AC inverter is used to convert DC voltage to AC voltage for the grid-tied application. It also eliminates reasonable harmonics in the system. This unidirectional DC–DC converter controls the fuel cell voltage and regulates the inverter input voltage. The DC current output of the SMES is converted to DC voltage by the current/voltage converter (I/V converter). Its main function is to keep constant DC voltage in the DC bus [25]. The detailed configuration of the combined operation of SMES and fuel cell for grid-tied application is provided in Figure 6.11.

The fuel cell cannot acknowledge current in the reverse direction, have a low yield voltage, and respond slowly to the step changes in load. Hence, the DC–DC converter is important to maintain the voltage profile and provide firm DC voltage. Also, the expansion of the DC–AC inverter has to be considered when designing the DC–DC converter. Different DC–DC converter configurations are available for fuel cell application, but the full bridge converter configuration provided in Figure 6.11 is used most frequently due to its ability to provide electrical isolation, and adaptability to high power transmission [26].

For almost all applications, DC–AC inverter is required for proper applications of the power conditioning of the fuel cell system. For three-phase inverter application, the conventional configuration of the three-phase PWM inverter with an output LC filter is show in Figure 6.11. The filter provides stable AC output with little ripples [27].

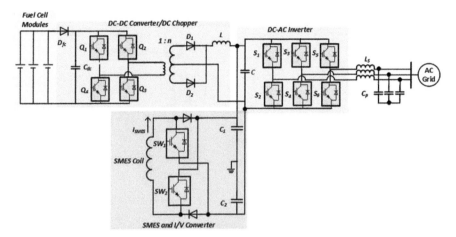

Figure 6.11   Detail scheme of the SMES–fuel cell combined operation [22, 25]

The SMES is also connected to the AC power grid through the bidirectional DC–AC inverter, which enables the re-energization to SMES at an allowable low power level. The primary function of the I/V converter of the SMES is to keep the constant voltage across the capacitors of the SMES. When the SMES capacitor voltages go below the setting value, the switch of the IGBTs is turned off to allow the SMES currents to flow into the capacitors and increase the capacitor voltage. On the other hand, when the voltage across the SMES capacitors goes above the setting value, the switches of the IGBTs are turned on to allow the SMES current circulate in the stead state mode.

6.5   Dynamic performance enhancement of power grids by combination of SMES and SVC

The SVC is a FACTS family device for fast and proper control of reactive power in the power system. The SVC was first developed in the early 1970s for the fast controlling of the power factor of changing loads, but nowadays they are used mainly for regulating the transmission voltage and improving the power quality situation. The SVC is connected in shunt with the power system, and it adjusts the voltage magnitude of the PCC by absorbing or producing reactive power [4, 28–30]. In SVC, a thyristor-controlled reactor (TCR) is used to control the reactive power by using proper control on the output of the equivalent susceptance function. Although SVC is primarily used for compensating the bus voltage by injecting or absorbing the reactive power, but it also has the capability of improving the transient stability of the power system. SVC can limit the damping of power system by utilizing the first swing accelerating and decelerating area of the generators load angle [4, 31]. The other applications of SVC are damping of low-frequency oscillations, damping

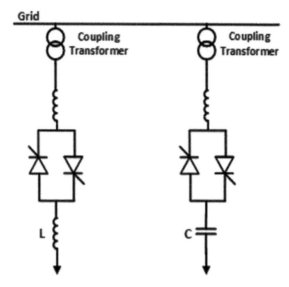

Figure 6.12   Schematic diagram of TCR and TSC for variable impedance-type SVC [33]

of sub synchronous frequency oscillations, control of dynamic overvoltages, etc. They are connected in parallel with the transmission line [32].

There are three basic types of SVC in the literature [32]. They are: (1) variable impedance-type SVC, (2) voltage source-type SVC, and (3) current source-type SVC.

Figure 6.12 shows the schematic diagram of TCR and thyristor-switched capacitor (TSC) for variable impedance-type SVC.

The variable impedance-type SVCs are mainly made of TCR in parallel with fixed capacitor (FC) or TSC. They mainly regulate the voltage in the transmission line. If the power system is capacitive, then TCRs absorb VARs from the system to maintain proper stability. On the other hand, if the power system is inductive in nature, then the FC or TSC automatically switched on and provides reactive power to the system for stabilizing the voltage. Both TCR and TSC are supplied from a step-down transformer, and the thyristors are controlled by controlling the firing angle range from 90 to 180 degrees [32]. TCR and TSC are made of power electronic devices.

To prevent the grid from the harmonics generated in the TCR and TSC, the step-down transformers are connected in star-delta formation. To suppress the remaining harmonics from the system, filters are used along with TCRs and TSCs. The FC–TCR-type SVC is effective in compensating the reactive power and controlling each phase voltage imbalances of the grid. It consists of the reactor, FC, and thyristor-controlled power switch. Figure 6.13 represents the FC-TCR-type SVC.

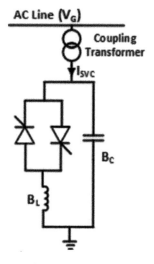

Figure 6.13   Fixed capacitor (FC) TCR-type SVC [33]

Figure 6.14 shows the schematic diagram of a voltage source-type SVC. Voltage regulation of the SVC is accomplished by varying the firing angle, α of the thyristors. The amount of voltage across the SVC can be adjusted to any desired value by varying the thyristors firing angle, α [34]. The magnitude of the grid voltage can further be compared with the reference voltage for better adjustment of the firing angle, α. The grid voltage at the PCC can be adjusted using the following control logic [31]

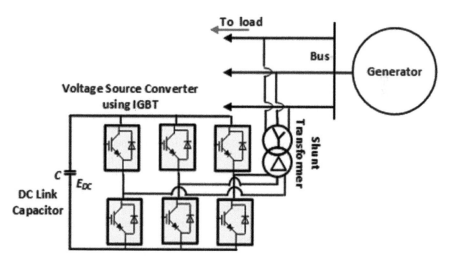

Figure 6.14   Voltage Source-type SVC.

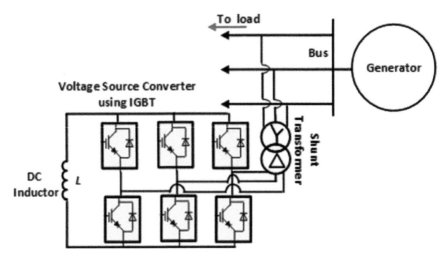

Figure 6.15　Current source-type SVC

$$V_{Ref} - V_G = j \frac{I_s}{B_{SVC}} \tag{6.1}$$

$$\text{where, } B_{SVC}(\alpha) = \frac{2\alpha - \sin(2\alpha) - \pi\left(2 - \frac{X_L}{X_C}\right)}{\pi X_L}$$

where V_{Ref} is the reference voltage, V_G is the grid voltage, I_s is the total SVC current, $B_{SVC}(\alpha)$ is the SVC susceptance determined with the control system gain as a function of firing angle α, X_L and X_C are, respectively, the inductive and capacitive reactance of the SVC, and α is the thyristor firing angle.

When the grid voltage, V_G, is less than the reference voltage, V_{Ref}, then SVC susceptance, B_{SVC}, is positive to inject reactive power into the power system. On the other hand, when the grid voltage, V_G, is more than the reference voltage, V_{Ref}, then SVC susceptance, B_{SVC}, is negative to absorb reactive power from the power system.

Voltage source-type SVC uses a voltage source converter (VSC) and a DC link capacitor in the terminal of VSC for proper controlling the reactive power of the system. The three-phase VSC is made of six IGBTs for proper controlling the voltage. The voltage from the grid is stored in the capacitor through VSC in normal operating condition. Moreover, in the event of disturbance, three-phase ac voltage is produced from the DC link capacitor voltage through VSC [32].

Figure 6.15 shows the schematic diagram of a current source-type SVC. Current source-type SVC uses a current source converter (CSC) with a DC inductor instead of a DC link capacitor in VSC. In the CSC, the DC current is kept constant with small ripples and thus it forms a current source working on the DC side. In a CSC, the direction of current always remains in the same direction [32].

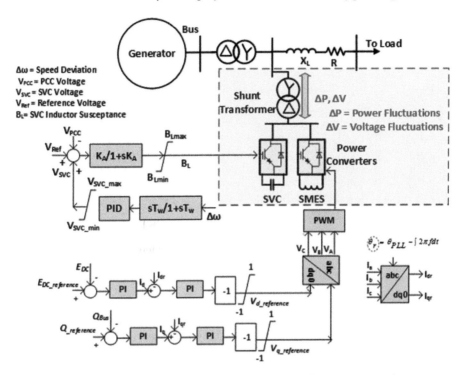

Figure 6.16   SMES and SVC coordinated operation with proper control systems [33, 36]

The SMES can control the active power as well as reactive power. So, it can simultaneously control the frequency and voltage. However, modern power system employs distributed energy resources (DER) for better operation, reliability, resilience, and to manage the distribution network capacity requirements. These DERs require enough grid forming and grid following inverters for their proper operation. These converters and inverters consume large amount of reactive power; therefore, rapid change of reactive power in power grid causes large voltage fluctuations and lower power factor. Therefore, it is important to expand the reactive power control region along with the active power. In order to reduce the voltage and frequency fluctuations, and stabilize the power instabilities, coordinated control and operation of the SMES and SVC can provide reliable load operation and dispatching [35]. Figure 6.16 shows the schematic diagram of the coordinated operation of SMES and SVC with their corresponding control structures.

The SMES and SVC connected in the load bus can absorb and supply the power deviation and voltage deviation, respectively, between the load and generator by means of compensating the weighted value of the load consumptions. In the configuration presented in Figure 6.17, when the load requires power, the SMES supplies the required power, and in the event of surplus power from the generator, it absorbs it to smooth the rapid changes in the load. Similarly, the SVC in the load bus is used to absorb/supply

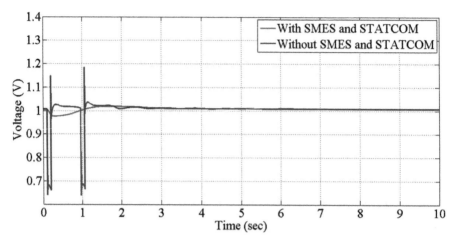

Figure 6.17   Voltage at the point of common coupling of the microgrid in different operating conditions

reactive power into the load depending on the difference between actual voltage and reference voltage. The operation of the SVC can be controlled by means of different linear and nonlinear controllers like, the PID controller, fuzzy logic controller, ANFIS controller, neural network-based controller, etc.

6.6   Dynamic performance enhancement of power grids by combination of SMES and STATCOM

The STATCOM is a shunt-connected FACTS family device that injects reactive current into the power system and controls the reactive power. Along with the reactive power control, the STATCOM improves the transient stability, power quality and damps the power system oscillations. The injected current by the STATCOM into the power system can be controlled by the power system voltage supplied through the power electronics-based variable voltage source. Unlike the SVC, the STATCOM does not employ any capacitor bank to produce reactive power. Rather in the STATCOM, the capacitor is used to produce constant DC voltage in order to provide the proper operation of the voltage source inverter (VSI). Figure 6.18 represents the reactive power generation scheme of the VSI-based STATCOM. The VSI produces a set of controllable output voltage. If the amplitude of the output voltage becomes more than the power system ac line voltage, then current flows through the leakage reactance from the converter to the ac system and the converter generates reactive power for the ac system. On the other hand, the amplitude of the output voltage becomes less than the power system ac line voltage, then current flows through the leakage reactance from the ac line to the converter and the converter absorbs reactive power [37]. If the voltage of the ac line and the converter output voltage amplitudes are equal, then the net reactive power exchange becomes zero.

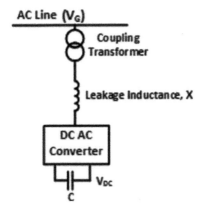

Figure 6.18   Voltage source inverter-based STATCOM connection scheme in power system

Many VSI topologies and configurations are adopted for designing the multi-level and multi-pulse STATCOM. There are applications of three-phase two-level 6 pulse bridge configuration, three-phase two-level 12 pulse bridge configuration, three-phase two-level 24 pulse bridge configuration, three-phase three-level 12 pulse configuration, and three-phase three-level 48 pulse configuration. The multilevel multi-phase configurations received wide attention due to their adaptability in high voltage and high power applications. Among them, the three-level topologies are most prominent because of their applicability in high voltage transmission systems. Furthermore, as gate turn-off thyristors (GTO) are commercially available for high power capacity and have the ability of triggering once per cycle, the GTO-based VSIs are extensively used for high power rating STATCOMs in high voltage transmission systems [38].

An STATCOM can only inject or absorb reactive power, limited in its degree of freedom and sustained action to support the power system. The STATCOM has only leading (capacitive) and lagging (inductive) mode of operation. The active power cannot be independently adjusted by only using the STATCOM. The addition of SMES along with the operation of the STATCOM provides simultaneous control of active and reactive power. Thus, they provide outstanding dynamic performance due to fast and independent exchange of active power and reactive power with the power system. Simultaneous operation of SMES and STATCOM can provide four different modes of operation: inductive with DC charge, inductive with DC discharge, capacitive with DC charge, and capacitive with DC discharge.

However, the independent shunt-connected structure of the SMES and STATCOM may not provide the optimal advantage of fast, coordinated, and independent operation. Therefore, the self-commutated VSI of the STATCOM can be utilized as the power conversion unit of the SMES. The SMES can dynamically exchange the active power with the power system by utilizing the SMES coil coupled with the DC–DC chopper. So, the DC–DC chopper acts as the interface between the STATCOM and the SMES and enables the bidirectional energy transfer

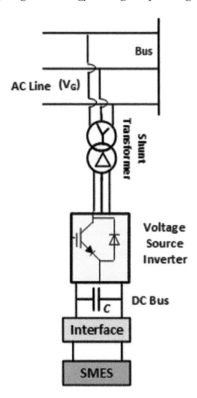

Figure 6.19   STATCOM and SMES integrated together through power
electronics interface [39, 40]

as represented by Figure 6.19. This arrangement can increase the overall performance and reduce the cost of the power electronics unit needed for the SMES.

The VSI is a DC to AC switching power converter created using GTO to generate balanced three-phase sinusoidal voltage at the fundamental frequency and
involves a phase control scheme suitable for high power transmission applications.
The details connection scheme of the STATCOM and SMES is represented in
Figure 6.20, which involves three-phase three-level 12 pulse VSI. This inverter is
referred as three level because of the three DC voltage levels: $+\dfrac{E_{dc}}{2}$, 0, and $-\dfrac{E_{dc}}{2}$.
The zero level (0) is obtained by connecting a clamping diode to the midpoint of
the DC capacitors. So, this structure gives an additional advantage of the levels
in the output voltage. On the other hand, the DC–DC chopper has three modes
of operation: charging, discharging, and standby. The DC–DC chopper works as
step-down converter during charging mode, as step-up converter during discharging mode. After the completion of SMES coil charging it works in the standby
mode [40].

The operation of the STATCOM can be controlled using different linear and
nonlinear control methodologies like PI controller, predictive controller, feedback

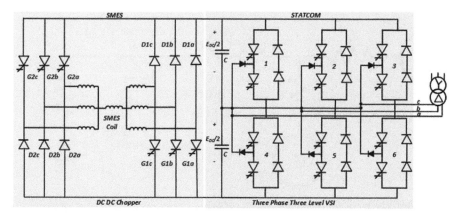

Figure 6.20   Detail connection scheme of the STATCOM, DC–DC chopper, and SMES [39, 40]

controller, fuzzy logic controller, neural network controller, etc. [40]. Coordinated operation of the SMES and STATCOM can be controlled using the external-, middle-, and internal-level control schemes. External-level controller is responsible for active-reactive power exchange between STATCOM and SMES, middle-level controller controls the active-reactive power exchange between the STATCOM–SMES and ac power system, and the internal-level controller is responsible for triggering and blocking operation of the VSI and DC–DC chopper.

6.7   Dynamic performance enhancement of power grids by combination of SMES and fault current limiters

Transient stability is the ability of the system to maintain stability after the occurrence of severe disturbance. This may result in large excursions of generator rotor angle and loss of system synchronism. The machines in the power system have to maintain synchronism after such disturbances. These disturbances may be in the form of sudden application of load, loss of generation, loss of large load, or a severe fault in the system. Instability is usually due to the insufficient synchronizing torque in the form of aperiodic angular separation. The equal area criterion can be used for quick prediction of the stability [41, 42]. According to equal area criterion, if the system is stable in the first swing, then the rotor angle after any disturbance reaches a maximum value. Also, the rotor angle oscillates about the final steady state value [41].

An SMES can be used to swiftly exchange the active and reactive powers with the power system to enhance the transient stability. However, the SMES cannot absorb enough energy during severe disturbance and faults since the bus voltage in the SMES terminal drops substantially. Moreover, large capacity of converter and energy storage of SMES are required to attain high stabilizing effect. Thus, to improve the control performance of the SMES, FCLs are applied along with the SMES. The operation of the SMES and FCL is coordinated where the FCL absorbs

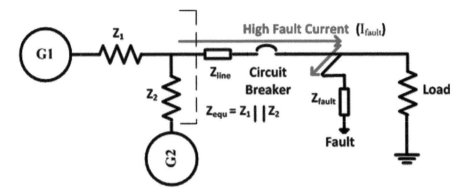

Figure 6.21   AC line to ground fault situation in a power system

the generator accelerating power during fault and supports the SMES to stabilize the power swing [43].

The wind generators are very vulnerable to grid faults as their stator windings are directly connected to the grid. When any grid fault happens in the wind generator system, the rotor side converter (RSC) and DC link capacitor, respectively, experience serious overcurrent and overvoltage, which may lead to the destruction of the converters. Moreover, when the system fault happens, the wind generator bus voltage drops significantly too. Therefore, individual solutions can only solve the problem by reducing the fault rotor current or increasing the terminal voltage of the wind generator [44]. The coordinated operation of SMES and FCL can solve the problem of high fault current and unsteady output power simultaneously [45, 46]. During the variable wind speed operation, the SMES can operate to smooth the active power of wind generator and improve the voltage sag situation while the FCL can operate to limit the high rotor current as well as fault current during fault condition in the grid [47, 48].

A fault in a power system creates temporary or permanent low impedance current paths between two circuit terminals. In power systems, a fault can happen for many reasons, such as insulation failures, birds or animals crossing the power lines, broken tree on the power lines, or even operating errors. Figure 6.21 represents a ground fault situation in a power system.

Although the impedance created by the line to ground faults are not predictable but they are small values. Furthermore, their value is way smaller than the impedances of the load. Therefore, in line to ground fault situations, the generator output voltages supply only the combined impedances from source (equivalent impedance), transmission line, and the fault point, which can be expressed as

$$I_{fault} = V_G/(Z_{equ} + Z_{line} + Z_{fault}) \qquad (6.2)$$

where V_G is the source voltage, Z_{equ} is the source equivalent impedance, Z_{line} is the line impedance, and Z_{fault} is the fault impedance.

Normally, the source impedance and the line impedances are kept low to improve the transmission efficiency. Because of the inertial of the generators, the

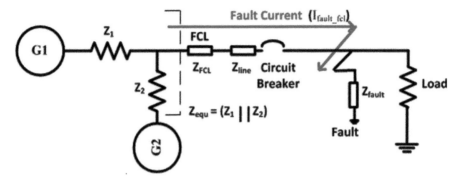

Figure 6.22   Increased impedance due to the presence of FCL

output voltage from the generators remains constant even in the fault situations. Therefore, when a fault situation arises in a power system, the high source voltage and low fault impedance generate high fault current, which feeds the fault point bypassing the load. Thus, the equipment, conductors, etc. situated between the generator and the fault point, experience high level of current and stress.

The fault currents generated are very destructive. Because sometimes the generated fault current is 20 times the normal operating currents. These high fault currents can generate massive heat and stress on the conductors, switchgear devices, transformers, and protective devices. If not properly handled, they can cause serious injuries, fires, and explosions too. Therefore, fault currents must be properly and seriously handled.

In (6.2), it is indicated that the fault current level can be reduced either by increasing the total impedance or by reducing the source voltage. However, the impedance increasing aspect is mostly adopted for FCL applications. Figure 6.22 represents the increased impedance effect in the presence of FCL.

The FCL inserts impedance only in the event of the fault and remains out of the circuit in the normal operating conditions. Therefore, it has no effect on the faulted circuit during normal operation. Therefore, in the faulty situation, as represented in Figure 6.22, the fault current passes through the power system can be represented as

$$I_{fault\_fcl} = V_G/(Z_{equ} + Z_{FCL} + Z_{line} + Z_{fault}) \qquad (6.3)$$

where Z_{FCL} is the impedance of the FCL.

By comparing (6.3) with (6.2) then it can be revealed,

$$I_{fault} > I_{fault\_fcl}$$

Thus, the FCL impedance dominates the total impedance of the power line during fault conditions and reduces the effect of high fault current.

Figure 6.23 shows the coordinated operation of the SMES and FCL, which consists of individual SMES unit as well as FCL unit shared by superconducting coil (SC)/SMES coil. The SMES unit consists of the VSC and the DC chopper, whereas the FCL unit consists of diode bridge rectifier.

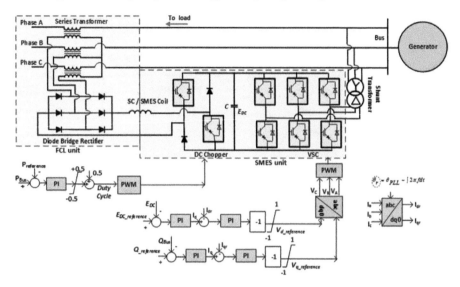

Figure 6.23   Configuration of the coordinated SMES and FCL unit [49]

During normal operation, the DC chopper controls the charging and discharging operation of the energy between the SC and the power grid [49]. In this situation, the rate of change of current in the SC as well as the induced voltage across it is almost zero. Furthermore, the equivalent impedance of the series transformer and the impedance of the SC via the bridge rectifier appear almost zero to the AC line. Different control mechanisms like the PI controller are used to decide about the charging and discharging operation of the SC. If the generated power is less than the reference power, then the SC discharges energy to keep the generated power up to the acceptable level. However, if the generated power is more than the reference power, then the IGBT switches and the diodes of the DC chopper are turned on and off to charge the SC. During no power difference situation, the SC current circulates across the IGBT switches and diodes to keep the SC energy to without transferring to or from the power grid.

On the other hand, during the fault situation, the rate of change of current in the SC with respect to time is high. Thus, the voltage induced across the SC becomes high and results in high equivalent SC impedance in the AC line. The VSC can be turned off to initiate the operation of the FCL unit using the PI controller using the line current of the power grid as the control input.

6.8   Case study

In this section, dynamic performance enhancement of the power grid by coordinated operation of the SMES and other control systems has been discussed by showing some examples emphasizing the role of different auxiliary devices along with the SMES in maintaining the grid dynamic performance. Different test systems have

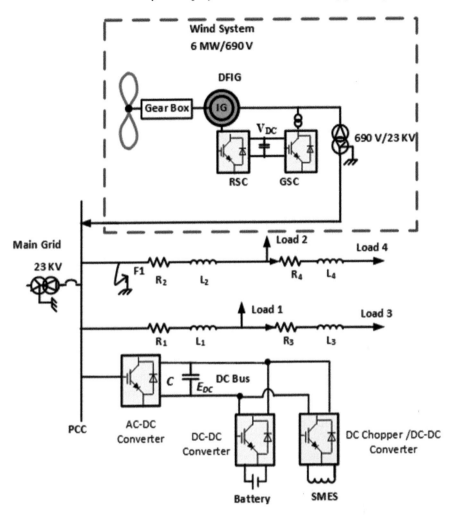

Figure 6.24   Microgrid with the SMES and BESS

been adopted to demonstrate the effectiveness of the SMES in coordinated operation with the BESS, STATCOM, SVC, and FCL in improving voltage profile, frequency smoothing, power balancing, and grid contingency support. These coordinated operations are adopted to contribute in oscillation damping during disturbance in the system.

Case study 1: dynamic performance improvement of power grids by combinations of SMES and battery energy storage

The microgrid system model shown in Figure 6.24 has been used for simulation in this case study. This microgrid consists of double circuit distribution lines, which are delivering power to loads. The microgrid can operate in standalone mode and grid-connected mode while connected to 23 KV distribution substation. A 6 MW doubly

Table 6.2   Wind generator, turbine, battery unit, SMES unit, and line parameter data

Characteristics	Value
Nominal power	1.5 MVA
Rated voltage	690 V
Distribution substation voltage	23 KV
Frequency	60 Hz
Stator to rotor turn ratio	0.35
Stator resistance (R_s)	0.027 pu
Stator leakage inductance (L_{ls})	0.18 pu (referred to stator)
Rotor resistance (R_r)	0.06 pu
Rotor leakage inductance (L_{lr})	0.016 pu (referred to stator)
Magnetizing inductance (L_m)	2.9 pu
DC link rated voltage	1150 V
Turbine inertia constant	4.32 s
Shaft spring constant	1.11 pu of nominal mechanical torque/rad
Positive sequence impedance	0.282676 + j 0.523848
Zero sequence impedance	0.553648 + j 1.18096
SMES unit	8 H, 600 V DC bus
Battery unit	1180 Ah, 600 V

fed induction generator (DFIG)-based wind farm is connected at the PCC of the microgrid through a step-up transformer. The microgrid line parameters and the wind generator parameters are presented in Table 6.2. Moreover, for achieving better power and energy controllability, the SMES and BESS are connected in the same PCC using power electronics interfaces like DC–DC converter and AC–DC converter. The wind farm local distribution line and the substation double circuit distribution lines have the circuit breakers. The 6 MW wind farm is built from four 1.5 MVA wind turbines. The detailed design and modeling of RSC and grid-side converter (GSC) for the DFIG-based wind generator are available in [50, 51].

In this case study, simulations have been performed by using the MATLAB/ Simulink software. Simulations have been carried out considering balanced (three-phase-to-ground (3LG)) permanent fault at location F1 on one of the microgrid distribution lines. The simulation time and time steps are considered as 50 seconds and 50 μs, respectively. For the permanent fault, it is considered that the circuit breakers in the system opens after 0.0833 seconds of the fault initiation and initiates a reclosing according to the regulations provided by the US transmission and distribution utilities, i.e., five cycles [52]. However, as the circuit breakers reclose during a permanent fault, then they reopen again after 0.0833 seconds of the reclosing time, which means an unsuccessful reclosing.

This case study determines the voltage and frequency responses at the PCC during 3LG fault at F1. Three operating conditions are considered: (1) without any SMES and BESS, (2) with BESS only, and (3) with SMES and BESS. Detail structure with proper control systems of the SMES and BESS is discussed in section 6.3.

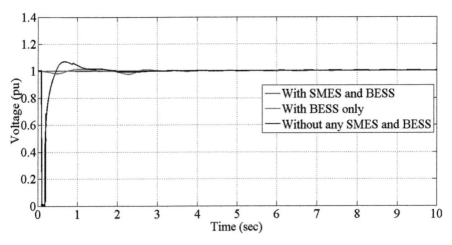

Figure 6.25   Voltage at the point of common coupling of the microgrid in different operating conditions

Figures 6.25 and 6.26 represent the voltage and frequency response at the PCC, respectively. These responses indicate that while the system experiences sudden disturbance, the combined operation of SMES and BESS stabilizes the system dynamics earlier. The dynamic responses of the wind generator are also stabilized in the presence of combined operation of SMES and BESS. The rotor speed response of the wind generator during the considered operating conditions is presented in Figure 6.27.

Case study 2: dynamic performance improvement of power grids by combinations of SMES and SVC

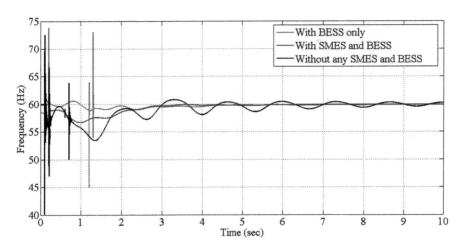

Figure 6.26   Frequency at the point of common coupling of the microgrid in different operating conditions

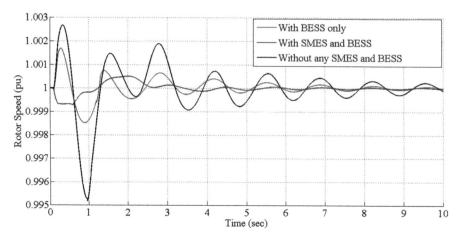

Figure 6.27   Rotor speed of the wind generator in different operating conditions

The microgrid system model shown in Figure 6.28 has been used for simulation in this case study. This microgrid system model has all the similar parameters like the previous case study model as represented in Figure 6.24. Moreover, for achieving better active and reactive power controllability, reliable load operation and dispatching, the SMES and SVC are connected in the PCC using power converters and shunt transformer. The wind farm local distribution line and the ac main grid, and double circuit distribution lines have the circuit breakers. The 6 MW wind farm is built from four individual 1.5 MVA wind turbines. The detailed design and model parameters are discussed in the previous case study. The SVC is rated for 2.5 MVAR with susceptance value of $B_c = 0.78$ pu, and the SMES coil is rated as 3 MW with coil rating of 0.64 H.

This case study determines the stability of the wind generator in the microgrid during 3LG fault at F1 at different time instances. Along with the wind generator, the PCC voltage stability is also studied. Two operating conditions are considered: (1) without any SMES and SVC, and (2) with SMES and SVC. Detail structure with proper control systems of the SMES and SVC is discussed in section 6.5. Figures 6.29 and 6.30 represent the power and DC link voltage of the wind generator, respectively. These responses indicate that during fault in the microgrid the combined operation of SMES and SVC can stabilize the distributed energy sources. The dynamic voltage response at the PCC during the considered operating conditions is also presented in Figure 6.31. These case study results indicate the effectiveness of the combined operation of the SMES and SVC in improving the dynamic stability of the overall power grid during transient instability.

Case study 3: dynamic performance improvement of power grids by combinations of SMES and STATCOM

The microgrid system model shown in Figure 6.32 has been used for simulation in this case study. This microgrid has a wind generator DER connected at the PCC. Wind generator parameters and the line parameters of the microgrid are provided

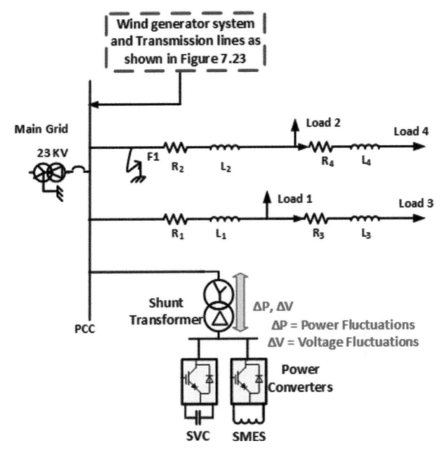

Figure 6.28   *Microgrid with the SMES and SVC*

in the previous section. For improving transient stability and power quality, damping the power system oscillations, controlling the active and reactive power, the SMES and STATCOM are connected in the PCC using VSI and shunt transformer. The capacitor of the STATCOM is used to produce constant DC voltage in order to provide the proper operation for the VSI. Circuit breakers in the wind farm local distribution line and the ac main grid are utilized to ensure the islanded and grid-connected operation of the microgrid. The 6 MW wind farm is built from four individual 1.5 MVA wind turbines. The 3 MW SMES coil is attached to a 2.5 MVAR STATCOM through DC–DC chopper and VSI at the PCC.

    This case study examines the overall stability of the microgrid. As a part of the case study, the wind generator's fault ride through ability is examined along with the stable load support ability of the microgrid. Two operating conditions are considered: (1) without any SMES and STATOM, and (2) with SMES and STATCOM. Detail structure with proper control systems of the SMES and STATCOM is discussed in section 6.6. Figures 6.33 and 6.34 represent the power and rotor speed of

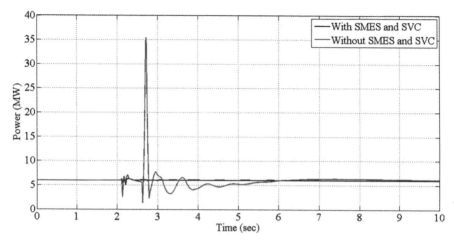

Figure 6.29   Wind generator power response for different operating condition

the wind generator, respectively. Figure 6.17 represents the voltage response at the PCC. These responses indicate the effectiveness of the combined operation of SMES and STATCOM in enhancing the fault ride through the ability of the wind generator. So, the combination of SMES and STATCOM can also improve the dynamic stability of the microgrid during transient instability.

Case study 4: dynamic performance improvement of power grids by combinations of SMES and fault current limiters

Detail modeling of this case study including the generators, SMES, and FCL is carried out extensively through the IEEE 39 bus system model [6]. Also, the wind generator model involving complete power electronic switching circuits, proper adaptive

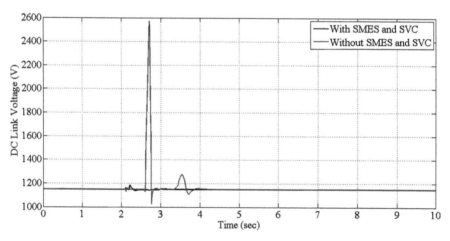

Figure 6.30   Wind generator DC link voltage response for different operating condition

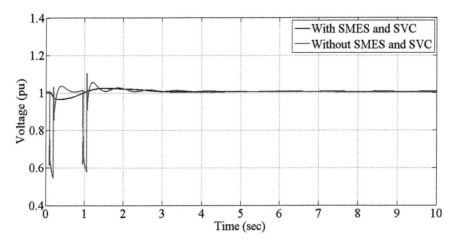

*Figure 6.31   Voltage at the point of common coupling of the microgrid in
different operating conditions*

control, and double circuit transmission line is adopted and integrated in the IEEE
39 bus power system [6]. The simulation design and results are carried out in the
MATLAB/Simulink environment. The simulation time and time steps are considered
as 20 seconds and 50 μs for getting accurate results. Two operating conditions are considered: (1) without any SMES and FCL, and (2) with SMES and FCL.

The IEEE 39 bus system is used in this work as a multi-machine system to
implement the combined operation of SMES and FCL for realistic response study.
The system consists of ten generators. There are 36 double circuit transmission lines
and 12 transformers in the system. Out of the 39 buses, 19 are the load buses. The
total load and generation of the system are 6098.1 and 6140.81 MW, respectively.
The system is represented in Figure 6.35. For each generator, the IEEE type 1 synchronous machine excitation and governor system have been used [53]. In order to
test response of the wind farm in a hybrid grid situation, a wind farm is connected
to bus 1 of the IEEE 39 bus system through an RSC, a DC link capacitor, a GSC,
three-phase step-up transformer, and double circuit transmission line as shown in
Figure 6.35. A simplified equivalent power grid with the presence of wind generator
is presented in Figure 6.36.

The parameters of the DFIG in wind farm and the drive trains are provided in
Table 6.3. In this case study, symmetrical (3LG) permanent fault considered to per-
sist for long time in the fault location F of the system as presented in Figure 6.35.
Fault in the double circuit transmission line of the wind generator is considered to
take place in different time instances. For the permanent fault, it is considered that
the circuit breakers in the system opens after 0.0833 seconds of the fault initiation
and initiates a reclosing according to the regulations provided by the US transmission and distribution utilities, i.e., five cycles [52]. However, as the circuit breakers
recloses during a permanent fault, then they reopen again after 0.0833 seconds of the
reclosing time, which means an unsuccessful reclosing.

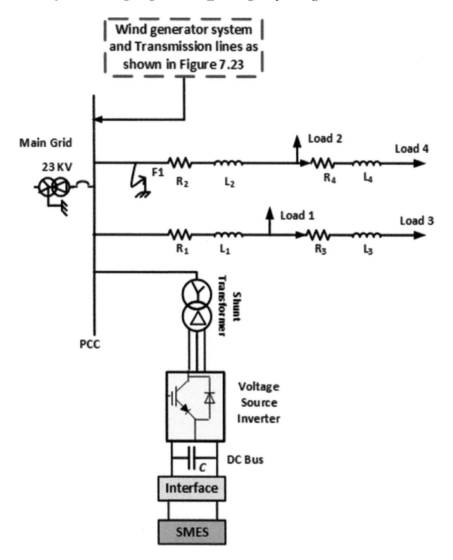

Figure 6.32   Microgrid with the SMES and STATCOM

The objective of utilizing FCL in this work is to minimize the high fault current, whereas the SMES improves the transient response as well as enhances the active power controllability of the overall grid. Based on these objectives, in this case study, we observed the current responses in different locations, power stability of the integrated wind generator, and the response of the rest of the system. Figures 6.37 and 6.38 represent the terminal current and voltage of the wind generator in different operating conditions. These results verify the effectiveness of FCL in reducing the high fault current and the ability of the SMES in improving the transient unstable voltage situation.

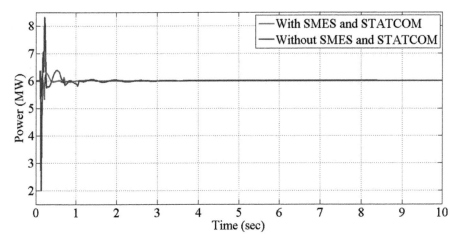

Figure 6.33   Wind generator power response for different operating condition

This case study further investigates the speed response of the wind generator. The speed gets high acceleration as the fault happens in the wind generator transmission lines. However, the high acceleration of the wind generator becomes stabilized in the presence of the SMES and FCL.

Further to understand the effect of transients on the rest of the system, this study also examines the load angle response of generator 1 in the IEEE 39 bus system. During fault in the system, the load angle of this synchronous generator becomes oscillatory in nature as presented in Figure 6.39. However, the combined operation of SMES and FCL stabilizes load angle response of the generator. The current passing through the FCL unit is also investigated for this study. As the FCL will be effective only for a very short duration, the current is observed for five cycles as

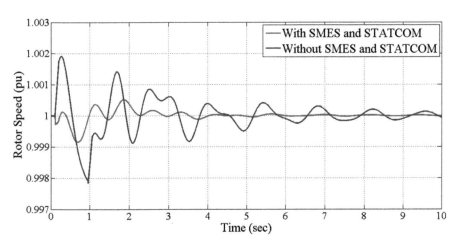

Figure 6.34   Rotor speed of the wind generator for different operating condition

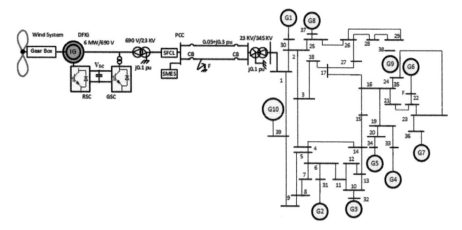

Figure 6.35   IEEE 39 bus system with integrated wind generator

represented in Figure 6.40. This current response also indicates the effectiveness of the combined operation of SMES and FCL in reducing the current magnitudes.

6.9   Cost effectiveness analysis

Every ESS has its own advantages and limitations. Ideal ESS application demands both high power and high energy capabilities. However, due to the limitations of individual ESSs capacities, it is necessary to build an HESS by combining more than one ESS. Therefore, the capital installation cost of any HESS is dominated by the energy-related (ESS-E) capital cost ($/€ per KW-h) and power-related (ESS-P) capital cost ($/€ per KW-h). Because, the high energy ESS (ESS-E) has less cost per KW-h, less life cycle, and less discharge capacities. On the other hand, the high power ESS (ESS-P) has higher cost per KW-h, longer lifetime, and high discharge capacities. Thus, HESS with less capital cost and high lifetime is most desirable in

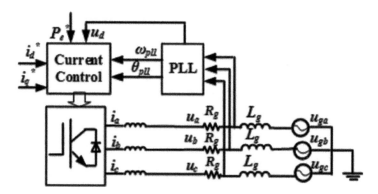

Figure 6.36   Simplified equivalent power grid in the presence of wind generator

Table 6.3   Doubly fed induction generator (DFIG) parameters

Parameters	Values	Parameters	Values
Rated power	6 MW	Stator resistance	0.005 pu
Frequency	60 Hz	d-axis reactance	0.092 pu
Wind speed	14 ms^{-1}	q-axis reactance	0.095 pu
Moment of inertia	0.8 Kg m^2	DC link capacitor	12,000 μF

order to minimize the capital installation cost, maintenance cost, operation cost, and replacement cost in order to improve the overall reliability of the ESS. Although SMES has high response speed, high power density, and operation density, there are several practical issues that hinder its extensive applications. The SMES incurs high costs due to superconducting components, cooling and cryogenic arrangements, and overall big size of the SMES unit. Therefore, the capital installation cost of SMES is very high compared to other available ESS. Among the high power density ESS, the per KW-h capital installation cost of the SMES is only comparable with the installation cost of super capacitor, which is €10,000–€20,000 per KW-h [7].

The capital installation cost of the BESS depends on the type of the energy storage device and its electrochemical properties. For lithium-ion BESS, the per KW-h unit installation cost ranges between €300 and €800. It is due to the high cost of the lithium and its recycling needs. The lead acid-type BESS require less cost for investment and its per KW-h unit installation cost ranges between €150 and €200. The sodium–sulfur (NaS)-type BESS is very costly due to its high production cost and recycling need for the sodium. However, it has high efficiency and high energy density. Its per KW-h unit installation cost ranges between €500 and €700 [7].

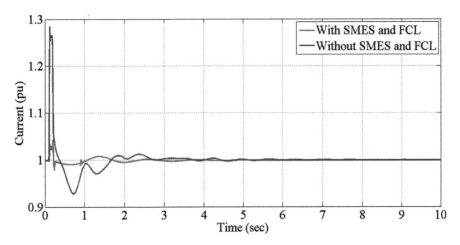

Figure 6.37   Terminal current of the wind generator

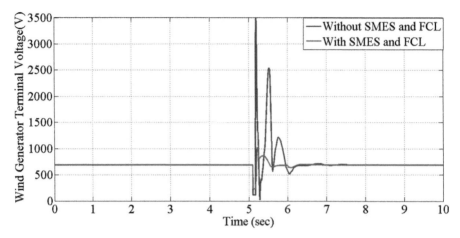

Figure 6.38   Terminal voltage of the wind generator

The overall commercial application of the fuel cell ESS for the standalone and grid-connected power system is yet to achieve the desired maturity. Biggest obstacles for its applications are the higher reaction time, moderate life cycle time, slow discharge rate, and less system efficiency. Any fuel cell application for standalone power system requires more than 40,000 working hours with not less than 10% decays [22]. The installation cost of the fuel cell ranges between €1,500 and €2,000 in per KW unit.

The exact price of the STATCOM and SVC depends on many factors like their power ratings, number of branches in which they are connected, initial installation cost, operation cost, etc. Wide range of deployment of the STATCOM and SVC in the power network is limited due to these factors and their costs. Installation cost is

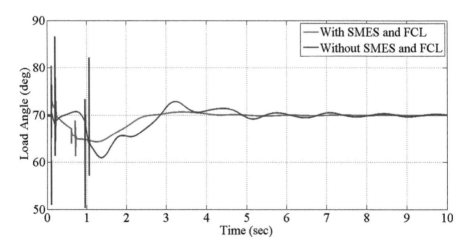

Figure 6.39   Load angle of generator 1 of the IEEE 39 bus system

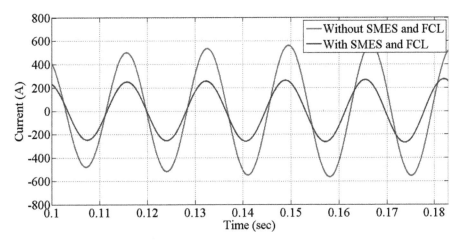

Figure 6.40   Current passing through the FCL unit

the one-time payment that includes the one-time purchase cost as well as the installation charges including the labor fees. However, the operation cost incurs over the lifetime operation of the STATCOM and SVC including the service and maintenance costs. Although an exact cost of the STATCOM and SVC is not available but based on the voltage levels their approximate costs are provided in different reports [54]. For the STATCOM, the price varies from \$60 to \$100 for per unit KVAR, and for the SVC, the price varies from \$40 to \$60 for per unit KVAR.

The price of the FCL also depends on the power ratings, type of the FCL, targeted voltage and current level, fault current level, current limiting requirements, type of the materials used, installation cost, and operation cost. The available FCL projects utilize the cooling system for the high-temperature superconductors (HTS), which incurs most of its costs [55, 56]. As an example, the project developed by the Zenergy Power for 138KV transmission class HTS FCL cost more than \$11 million [55].

Table 6.4 summarizes the characteristics of the discussed HESSs by analyzing their performances in different categories.

6.10   Conclusion

This chapter discusses the combined operation of SMES and other auxiliary devices in enhancing the dynamic performance of the power grid. Initially, the concept of HESSs has been discussed. The advantages of forming HESSs, different combinations of energy storage devices, and sharing their mutual advantages in improving the power grid dynamic stability have been discussed in the beginning of this chapter. Then the impact of combined operation of SMES–BESS, SMES–fuel cell, SMES–SVC, SMES–STATCOM, and finally SMES–FCL has been discussed one after another. These discussions involve the advantages of their combined operation, details about their structure, power electronics interface, primary control schemes,

Table 6.4 Summaries of the different HESSs characteristics

Type of the HESS	Control ability	Installation cost	Reaction time	Lifecycle	Application
SMES and battery	SMES – high power density; Battery – high energy density	SMES – $10,000–$22,000 per KW-h; Battery – $300–$1000 per KW-h	SMES – less than 10 ms; Battery – less than 5 ms	SMES – 10–20 years; Battery – 5–20 years	SMES – power quality, stability; Battery – peak load shaving, voltage stability
SMES and fuel cell	SMES – high power density; Fuel cell – high energy density	SMES – $10,000–$22,000 per KW-h; Fuel cell – $1500–$2200 per KW-h unit	SMES – less than 10 ms; Fuel cell – less than 2 minutes	SMES – 10–20 years; Fuel cell – 5–10 years	SMES – power quality, stability; Fuel cell – peak load shaving, load shifting, stability
SMES and SVC	SMES – active power and reactive power; SVC - reactive power	SMES – $11,000–$22,000 per KW-h; SVC – $40–$60 per unit KVAR	SMES – less than 10 ms; SVC – less than 20 ms	SMES – 10–20 years; SVC – 5–15 years	SMES – power quality, stability; SVC – voltage stability, improve transmission capacity
SMES and STATCOM	SMES – active power and reactive power; STATCOM – reactive power	SMES – $11,000–$22,000 per KW-h; STATCOM – $60–$100 per unit KVAR	SMES – less than 10 ms; STATCOM – less than 10 ms	SMES – 10–20 years; STATCOM – 5–15 years	SMES – power quality, stability; STATCOM – voltage stability, improve transmission capacity
SMES and FCL	SMES – active power and reactive power; FCL – active power	SMES – $11,000-$22,000 per KW-h; FCL – More than $8 million for transmission voltage-level FCL	SMES – less than 10 ms; FCL – less than 3 minutes	SMES – 10–20 years; FCL – 5–15 years	SMES – power quality, stability; FCL – transient stability, active power improvement

and DC–DC converter control schemes. Then the impact of these combinational operations on the power grid's dynamic stability during disturbance conditions has been analyzed. Power grid and microgrid test system models are developed in the MATLAB/Simulink software platform to carry out the analysis and detailed system responses have been provided at the end of the chapter.

References

[1] Colak I., Sagiroglu S., Fulli G., Yesilbudak M., Covrig C.-F. 'A survey on the critical issues in smart grid technologies'. *Renewable and Sustainable Energy Reviews*. 2016, vol. 54, pp. 396–405.

[2] Ali M.H., Dougal R.A. 'Comparison of SMES and SFCL for transient stability enhancement of wind generator system'. *2010 IEEE Energy Conversion Congress and Exposition (ECCE)*; 2010. pp. 3382–87.

[3] Ali M.H., Wu B., Dougal R.A. 'An overview of SMES applications in power and energy systems'. *IEEE Transactions on Sustainable Energy*. 2010, vol. 1(1), pp. 38–47.

[4] Hingorani N.G., Gyugyi L., El-Hawary M. 'Understanding facts: concepts and technology of flexible AC transmission systems'. *IEEE Press New York*. 2000.

[5] Farrokhabadi M., König S., Cañizares C.A., Bhattacharya K., Leibfried T. 'Battery energy storage system models for microgrid stability analysis and dynamic simulation'. *IEEE Transactions on Power Systems*. 2017, vol. 33(2), pp. 2301–12.

[6] Sadi M.A.H., AbuHussein A., Shoeb M.A. 'Transient performance improvement of power systems using fuzzy logic controlled capacitive-bridge type fault current limiter'. *IEEE Transactions on Power Systems*. 2020, vol. 36(1), pp. 323–35.

[7] Babu T.S., Vasudevan K.R., Ramachandaramurthy V.K., Sani S.B., Chemud S., Lajim R.M. A comprehensive review of hybrid energy storage systems: converter topologies, control strategies and future prospects. *IEEE Access: Practical Innovations, Open Solutions*. vol. 8, pp. 148702–21.n.d.

[8] Chong L.W., Wong Y.W., Rajkumar R.K., Rajkumar R.K., Isa D. 'Hybrid energy storage systems and control strategies for stand-alone renewable energy power systems'. *Renewable and Sustainable Energy Reviews*. 2016, vol. 66, pp. 174–89.

[9] Hajiaghasi S., Salemnia A., Hamzeh M. 'Hybrid energy storage system for microgrids applications: a review'. *Journal of Energy Storage*. 2016, vol. 21, pp. 543–70.

[10] Jing W., Hung Lai C., Wong S.H.W., Wong M.L.D. 'Battery-supercapacitor hybrid energy storage system in standalone DC microgrids: are view'. *IET Renewable Power Generation*. 2017, vol. 11(4), pp. 461–69. Available from https://onlinelibrary.wiley.com/toc/17521424/11/4

[11]   Jing W., Lai C.H., Wong W.S.H., Wong M.L.D. 'A comprehensive study of battery-supercapacitor hybrid energy storage system for standalone pv power system in rural electrification'. *Applied Energy*. 2018, vol. 224, pp. 340–56.

[12]   Nehrir M.H., Wang C., Strunz K. 'A review of hybrid renewable/alternative energy systems for electric power generation: configurations, control, and applications'. *IEEE Transactions on Sustainable Energy*. 2018, vol. 2(4), pp. 392–403.

[13]   Akinyele D.O., Rayudu R.K. 'Review of energy storage technologies for sustainable power networks'. *Sustainable Energy Technologies and Assessments*. 2014, vol. 8, pp. 74–91.

[14]   Chen L., Chen H., Li Y., *et al.* 'SMES-battery energy storage system for the stabilization of a photovoltaic-based microgrid'. *IEEE Transactions on Applied Superconductivity*. vol. 28(4), pp. 1–7.n.d.

[15]   Wang G., Ciobotaru M., Agelidis V.G. 'Power smoothing of large solar pv plant using hybrid energy storage'. *IEEE Transactions on Sustainable Energy*. 2014, vol. 5(3), pp. 834–42.

[16]   Sun Q., Xing D., Alafnan H., Pei X., Zhang M., Yuan W. 'Design and test of a new two-stage control scheme for SMES-battery hybrid energy storage systems for microgrid applications'. *Applied Energy*. 2019, vol. 253, p. 113529.

[17]   Shim J.W., Cho Y., Kim S.-J., Min S.W., Hur K. 'Synergistic control of SMES and battery energy storage for enabling dispatchability of renewable energy sources'. *IEEE Transactions on Applied Superconductivity*. 2013, vol. 23(3), pp. 5701205–5701205.

[18]   Li J., Yang Q., Robinson F., Liang F., Zhang M., Yuan W. 'Design and test of a new droop control algorithm for a SMES/battery hybrid energy storage system'. *Energy*. 2017, vol. 118, pp. 1110–22.

[19]   Lin X., Lei Y. 'Coordinated control strategies for SMES-battery hybrid energy storage systems'. *IEEE Access*. 2017, vol. 5, pp. 23452–65.

[20]   Khodadoost Arani A.A., B. Gharehpetian G., Abedi M. 'Review on energy storage systems control methods in microgrids'. *International Journal of Electrical Power & Energy Systems*. 2019, vol. 107, pp. 745–57.

[21]   Cansiz A., Faydaci C., Qureshi M.T., Usta O., McGuiness D.T. 'Integration of a SMES–battery-based hybrid energy storage system into microgrids'. *Journal of Superconductivity and Novel Magnetism*. 2018, vol. 31(5), pp. 1449–57.

[22]   Das V., Padmanaban S., Venkitusamy K., Selvamuthukumaran R., Blaabjerg F., Siano P. 'Recent advances and challenges of fuel cell based power system architectures and control – a review'. *Renewable and Sustainable Energy Reviews*. 2017, vol. 73, pp. 10–18.

[23]   İnci M., Türksoy Ö. 'Review of fuel cells to grid interface: configurations, technical challenges and trends'. *Journal of Cleaner Production*. 2018, vol. 213, pp. 1353–70.

[24]   Akinyele D., Olabode E., Amole A. 'Review of fuel cell technologies and applications for sustainable microgrid systems'. *Inventions*. 2020, vol. 5(3), p. 42.

[25] Yu X., Starke M.R., Tolbert L.M., Ozpineci B. 'Fuel cell power conditioning for electric power applications: a summary'. *IET Electric Power Applications.* 2007, vol. 1(5), p. 643.

[26] İnci M. 'Active/reactive energy control scheme for grid-connected fuel cell system with local inductive loads'. *Energy.* 2020, vol. 197, 117191.

[27] Monai T., Takano I., Nishikawa H., Sawada Y. 'Response characteristics and operating methods of new type dispersed power supply system using photovoltaic fuel cell and SMES'. *in IEEE Power Engineering Society Summer Meeting*; Chicago, IL, IEEE, 2002. pp. 874–79.

[28] Sadi M.A.H., Hasan Ali M. 'Combined operation of SVC and optimal reclosing of circuit breakers for power system transient stability enhancement'. *Electric Power Systems Research.* 2014, vol. 106, pp. 241–48.

[29] Haque M.H. 'Improvement of first swing stability limit by utilizing full benefit of shunt facts devices'. *IEEE Transactions on Power Systems.* 2004, vol. 19(4), pp. 1894–902.

[30] Haque M.H. 'Evaluation of first swing stability of a large power system with various facts devices'. *IEEE Transactions on Power Systems.* 2008, vol. 23(3), pp. 1144–51.

[31] Boynuegri A.R., Vural B., Tascikaraoglu A., Uzunoglu M., Yumurtacı R. 'Voltage regulation capability of a prototype static var compensator for wind applications'. *Applied Energy.* 2012, vol. 93, pp. 422–31.

[32] Padiyar R.K. Power system dynamics stability and control. John Wiley & Sons Pte Ltd, New York; 1996.

[33] Niiyama K., Yagai T., Tsuda M., Hamajima T. 'Design of power control system using SMES and SVC for fusion power plant'. *Journal of Physics.* 2008, vol. 97, p. 012271.

[34] Naderipour A., Abdul-Malek Z., Heidari Gandoman F, *et al.* 'Optimal designing of static var compensator to improve voltage profile of power system using fuzzy logic control'. *Energy.* 2020, vol. 192, 116665.

[35] Zargar M.Y., Lone S.A. 'Voltage and frequency control of a hybrid wind-diesel system using SVC and predictively controlled SMES'. *2017 6th International Conference on Computer Applications in Electrical Engineering-Recent Advances (CERA)*; Roorkee, 2017. pp. 25–30.

[36] Wang L., Truong D.-N. 'Stability enhancement of a power system with a PMSG-based and a DFIG-based offshore wind farm using a SVC with an adaptive-network-based fuzzy inference system'. *IEEE Transactions on Industrial Electronics.* 2012, vol. 60(7), pp. 2799–807.

[37] Anju M., Rajasekaran R. Presented at 2013 International Conference on Computer Communication and Informatics (ICCCI). Coimbatore, Tamil Nadu, India.

[38] Li H., Cartes D., Steurer M., Tang H. 'Control design of STATCOM with superconductive magnetic energy storage'. *IEEE Transactions on Appiled Superconductivity.* 2005, vol. 15(2), pp. 1883–86.

[39] Molina M.G., Mercado P.E. 'Comparative evaluation of performance of a STATCOM and SSSC both integrated with SMES for controlling the

power system frequency'. Presented at 2004 IEEE/PES Transmission and Distribution Conference and Exposition; Latin America.

[40]   Al-Haddad K., Saha R., Chandra A., Singh B. 'Static synchronous compensators (STATCOM): a review'. *IET Power Electronics*. 2009, vol. 2(4), pp. 297–324.

[41]   Kundur P. Power System Stability and Control. New York, USA: McGrwa-Hill; 1994.

[42]   Kundur P. 'Definition and classification of power system stability IEEE/CIGRE joint Task force on stability terms and definitions'. *IEEE Transactions on Power Systems*. 2004, vol. 19(3), pp. 1387–401.

[43]   Ngamroo I., Vachirasricirikul S. 'Coordinated control of optimized SFCL and SMES for improvement of power system transient stability'. *IEEE Transactions on Applied Superconductivity*. 2011, vol. 22(3), pp. 5600805–5600805.

[44]   Xiao X.-Y., Yang R.-H., Chen X.-Y., Zheng Z.-X. 'Integrated DFIG protection with a modified SMES-FCL under symmetrical and asymmetrical faults'. *IEEE Transactions on Applied Superconductivity*. 2018, vol. 28(4), pp. 1–6.

[45]   Elshiekh M.E., Mansour D.-E.A., Zhang M., Yuan W., Wang H., Xie M. 'New technique for using SMES to limit fault currents in wind farm power systems'. *IEEE Transactions on Applied Superconductivity*. 2018, vol. 28(4), pp. 1–5.

[46]   Chen L., Chen H., Yang J, *et al.* 'Coordinated control of superconducting fault current limiter and superconducting magnetic energy storage for transient performance enhancement of grid-connected photovoltaic generation system'. *Energies*. 2017, vol. 10(1), p. 56.

[47]   Ngamroo I., Karaipoom T. 'Improving low-voltage ride-through performance and alleviating power fluctuation of DFIG wind turbine in DC microgrid by optimal SMES with fault current limiting function'. *IEEE Transactions on Applied Superconductivity*. 2018, vol. 24(5), pp. 1–5.

[48]   Xiao X.-Y., Yang R.-H., Chen X.-Y., Zheng Z.-X., Li C.-S. 'Enhancing fault ride-through capability of DFIG with modified SMES-FCL and RSC control'. *IET Generation, Transmission & Distribution*. 2018, vol. 12(1), pp. 258–66. Available from https://onlinelibrary.wiley.com/toc/17518695/12/1

[49]   Ngamroo I. 'Optimization of SMES-FCL for augmenting FRT performance and smoothing output power of grid-connected DFIG wind turbine'. *IEEE Transactions on Applied Superconductivity*. 2016, vol. 26(7), pp. 1–5.

[50]   Boukhezzar B., Siguerdidjane H. 'Nonlinear control of a variable-speed wind turbine using a two-mass model'. *IEEE Transactions on Energy Conversion*. 2011, vol. 26(1), pp. 149–62.

[51]   Okedu K.E., Muyeen S.M., Takahashi R., Tamura J. 'Wind farms fault ride through using DFIG with new protection scheme'. *IEEE Transactions on Sustainable Energy*. 2012, vol. 3(2), pp. 242–54.

[52]   Committee I.P.S.R. IEEE guide for automatic reclosing of line circuit breakers for AC distribution and transmission lines. USA: IEEE Power Engineering Society; 2003.

[53]   Sauer P.W., Pai M.A. Power system dynamics and stability. Saddle River, NJ: Prentice Hall Upper; 1998.

[54]   Milanovic J.V., Zhang Y. Global minimization of financial losses due to voltage sags with facts based devices. *IEEE Transactions on Power Delivery.* 2004, vol. 25(1), pp. 298–306.

[55]   Moriconi F., De La Rosa F., Darmann F., Nelson A., Masur L. 'Development and deployment of saturated-core fault current limiters in distribution and transmission substations'. *IEEE Transactions on Applied Superconductivity.* 2011, vol. 21(3), pp. 1288–93.

[56]   Llambes J., Weber C., Hazelton D. 'Testing and demonstration results for the transmission-level (138 Kv) 2G superconducting fault current limiter at superpower'. *In Applied Superconducting Conf Chicago, Illinois, USA.* 2008.

Chapter 7

Artificial intelligent controllers for SMES for transient stability enhancement

P. Mukherjee and V.V. Rao

7.1 Introduction

Artificial intelligence (AI) is an extensive area of computer science combined with robust data set concerned to represent human intelligence for building smart machines to solve complicated problems of various technological fields. The systems are enabled to perform logical and rational intelligence to think or make action like human. The modern systems that perform based on AI are autonomous vehicle, conversational bots, spam filters for e-mails, robo-advisors, recommendations engines like Netflix, and smart assistants like, Alexa, Siri.

To control a system, a designer developing a conventional control system creates a mathematical model of the system with some assumptions like time-invariance, linearity, etc. This model involves all the dynamics of the plant that influences its control. In case of developing an artificial intelligent controller (AIC) system, the AIC conceptually is modelled based on the behaviour of the system given as inputs. The designer does not require to know the internal dynamics of the system to be controlled. Therefore, designing AIC for very complex plant is possible.

In recent years, AICs are drawing the interest of researchers as the conventional mathematical controllers are not well suited with uncertain non-linear systems. By contrast, AIC can model the qualitative knowledge and reasoning processes of human intelligence without using accurate quantitative analyses. As a result, the controlling part of the system becomes advanced in operations.

An efficient controller of superconducting magnetic energy storage (SMES) is very important for solving various issues of power systems. Effectivity of SMES system in the power systems has been found in various areas like transient stability, voltage sag and swell, voltage stability, load frequency control, primary frequency stability during fault, solving the issues with renewable energy integration, etc. An appropriately designed AIC can enhance the performance of SMES to the pre-eminent level for different system parameters and instabilities.

This chapter deals with the design, operation and application of three AICs for controlling SMES operation in power grids. An overview is given on the controlling methods of these AICs. Based on their controlling methodology, design procedures of these AICs for SMES are explained. The design of AICs is primarily focused on the control of charging and discharging of SMES in response of deviation in variables in power grids. At the end of this chapter, a comparative case study is done by applying the three AICs separately to SMES with constant energy storage capacity and improving the stability of a wind generator-integrated power system.

7.2   AI methods

The three widely used AICs for SMES are fuzzy logic controller (FLC), artificial neural network controller (ANNC) and adaptive neuro fuzzy interface system controller (ANFISC). FLC is designed based on the imprecise thinking of human being. In FLC, the input and output of the system to be controlled are described by the collection of fuzzy rules like IF-THEN, etc. For its robustness and ease of modification, it is used in various applications in power systems like stabilization and power quality improvement. ANNC is a controlling tool that can be trained as per system requirement. It learns from the experience and past data examples. ANNC is used for various problems like pattern recognition, prediction, optimization, associative memory and control. ANFISC integrates the knowledge base training process of neural network (NN) into the fuzzy interface to learn from the available data set, in order to compute the membership function (MF) parameters. This helps FLC to trail the given input/output data. This is a hybrid intelligent strategy that merges self-learning capacity of ANNCs with decision-making ability of FLC to develop a competent control tool. The objective of training ANFISC model is to minimize the error between the actual target and output values of FIS process. FLC does not necessitate the acquaintance on the main system to initiate operation. The general comparison of structure, methodology and functionality of these controllers is given in Table 7.1. The most excellent AIC can be designed to control SMES for power system applications by comparing performance index of different controllers to solve the power system issues.

7.3   Fuzzy logic-based control of SMES for power grids

Fuzzy logic control is employed in numerous applications of consumer products. Some of the examples that use FLC include washing machines, air conditioner, vacuum cleaner, unmanned helicopter, anti-braking system in vehicles, control on traffic lights, subway system control, large economic systems, etc. The application of FLC is not only limited to industrial process control but also covers biomedical instrumentation and securities. In comparison with classical controller, FLC has been best applied in solving complex and poorly defined problems, irrespective of proper information about their fundamental dynamical equations. Although other AICs can operate just like FLC in many cases, fuzzy logic has the advantage that the problem can be solved in terms of human thinking and expertise and the controller can be designed based on human experience. Therefore, FLC can be designed with proper understanding and its effectiveness can be enhanced according to the desire.

Table 7.1   *Comparison of AI controllers for SMES*

AIC	Advantages	Disadvantages
ANNC	Adapt uncertainties	Complex network structure
	Self-tuning capacity	Hard to train
	Self-learning capacity	Impossible interpretation of functionality
	Generalization capacity	
	Robustness in relation to disturbance	
FLC	Adapt to uncertainties	Large no of parameters
	Flexible	Large no of options
	Fast computation	Incapable to generalize
	Decision-making ability	Robustness is limited to possible disturbances
	Inherent programming capability	Needs prior knowledge for rule discovery
	Simple network structure	
	Easy to interpret	
ANFISC	Adapt to uncertainties	
	Interpretable	
	Self-learning capacity	
	Self-tuning	
	Fast computation	
	Decision-making ability	
	Robustness in relation to disturbance	

The most challenging part of FLC is designing fuzzy set theory, deriving fuzzy reasoning and developing fuzzy logic. Fuzzy reasoning is an inference process that derives decisions from a set of fuzzy IF-THEN rules and subjected to instabilities of power system. Fuzzy set theory deals with problems relating to indistinct, subjective and inaccurate judgments. It can compute the linguistic facet of given data and find preferences on individual data for decision-making. Fuzzy logic is a method of computing based on 'degrees of truth' than Boolean representation of 'true or false' (logic 1 or 0). Fuzzy set theory is derived from Fuzzy logic.

SMES absorbs/delivers active power from/to the power system. Controllers are mostly used to control the converter and chopper circuits of SMES. Active power is controlled to level the output power, whereas reactive power is controlled to regulate the voltage profile of power system. The power drawn from stored energy of SMES coil should have a limit to avoid heating and to prevent loss of its superconductivity and coil burning. The overcharge and deep discharge process of SMES should also be controlled in accordance with these limits [1,2].

Fuzzy logic-based control of SMES can solve the stability issues of power grid due to both balanced and unbalanced disturbances [3] or can resolve the problem of unsuccessful reclosing of circuit breaker followed by the initiation of fault [4] or can

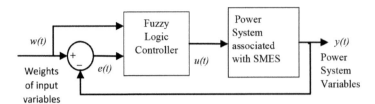

Figure 7.1   Fuzzy logic control system for SMES associated with power system

stabilize the system voltage in addition with static compensator [5]. Furthermore, it is cost-effective as it handles disturbances with reduced size of SMES [6] and is also better than static non-linear controller. The primary frequency can be effectively maintained by using only fuzzy-controlled SMES by supplying or absorbing extra power to or from power system for a sudden increase or decrease of frequency. Fuzzy-controlled SMES has a superior performance to improve the output power profile of wind power generating systems (WPGS) to grid. The SMES can stabilize the direct current (DC)-link voltage and smooth the output power simultaneously during normal condition. The back electromotive force induced in rotor due to the occurrence of fault in the connected system generally becomes several times higher than the rated RSC voltage and hence the RSC loses control over the rotor current [7]. The over current and electromagnetic torque oscillations in RSC can be eliminated by FLC-controlled SMES.

Figure 7.1 shows the block diagram of FLC for the power system, which includes the ancillary systems like SMES. $y(t)$ represents the power system variables, $w(t)$ represents the input weights of fuzzy controller, $e(t)$ is the error. The FLC is the combination of feedforward and feedback controller. $u(t)$ represents the output signal from FLC.

The deviation in power system variable followed by any disturbances is sensed by the controller as an error signal. The difference between the reference value of the variable and the deviated value produces the error signal. To adjust the error signal for obtaining stability in power system variables, FLC is added in the closed-loop control system. FLC mathematically analyses analogue input values of error signals in terms of logical variables that accept continuous values between 0 and 1. Therefore, the cost-effective, robust, customizable development of FLC replicate, the human deductive thinking for the control of the SMES, is possible to improve the power system reliability and efficiency. However, effective FLC development needs lots of data and regular updating of rules.

The block diagram of fuzzy logic control is shown in Figure 7.2. The error signals from power system and SMES in the form of crisp value enter into the fuzzy logic system. The crisp variables are then fuzzified into linguistic variables with the help of MFs. The fuzzy rules are set by the developer as per the desired output from SMES for a given issue of a power system. Fuzzy interface understands the rule for the fuzzified variables in the form of MFs and produces

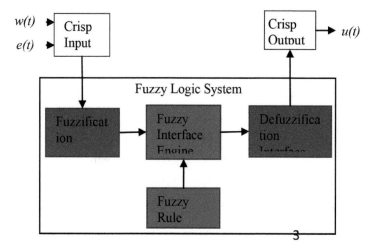

Figure 7.2   Block diagram of fuzzy logic controller

the output signal. The output signal from fuzzy interface is then converted into crisp value from fuzzy form by defuzzification interface. These crisp values are actually the control signals for converter circuits of SMES. Figure 7.3 shows a concept of universe of discourse and crisp set at somewhere in the universe. Figure 7.3a represents classical set with a certain boundary with an unambiguous line. Point A_1 is a member of crisp set and B_1 is definitely not a member of the set. Figure 7.3b shows a fuzzy set that has an ambiguous and shaded boundary. The point A_2 is a full member in the fuzzy set. The pointy B_2 is not a fuzzy member. Point C_2 is on the boundary of fuzzy set and represents the ambiguous membership. The complete membership (A_1 and A_2) is represented by 1 and no membership (B_1 and B_2) in a set is represented by 0. The membership of C_2 is somewhere between 0 and 1.

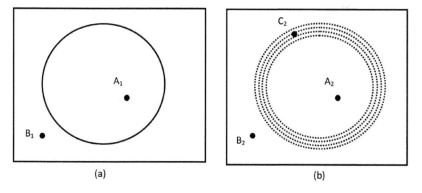

Figure 7.3   Notation of crisp membership (a) crisp set, (b) fuzzy set

If V is the signal or an input vector signal measured from the power system or SMES followed by any disturbance and m is the MF mapping the interval $[0,1]$ of the real signal V, then the fuzzy set can be defined as (V, m)

$$m: V \rightarrow [0, 1] \tag{7.1}$$

The mathematical expression for the fuzzy set is A when the universe of discourse, X, is discrete and finite

$$A = \left\{ \frac{\mu_A(x_1)}{x_1} + \frac{\mu_A(x_2)}{x_2} + ... \right\} = \left\{ \sum_i \frac{\mu_A(x_i)}{x_i} \right\} \tag{7.2}$$

When the universe X is continuous and finite, the fuzzy set A can be expressed as

$$A = \left\{ \int \frac{\mu_A(x)}{x} \right\} \tag{7.3}$$

Fuzzification of a signal X from the grid is represented by a combination of fuzzy sets s_i, $i = 1, ..., k$. A fuzzy set $s = (V, m)$. An arbitrary element $x \in X$ belongs to the fuzzy set s_i with degree $d_i = f_i(x)$, $i = 1, ..., k$. The fuzzy logic defines the rules governing the operators intersection and union of fuzzy sets. If there are two fuzzy sets for the same signal X: $s_1 = (X, f_1)$ and $s_2 = (X, f_2)$, then the arbitrary element belongs to either union $s_1 \cup s_2$ of the two fuzzy sets s_1 and s_2 with degree $d = max(f_1(x), f_2(x))$ or intersection $s_1 \cap s_2$ of two fuzzy sets s_1 and s_2 with degree $d = min(f_1(x), f_2(x))$. A fuzzy variable V is an ordered pair (s, d), where s is a fuzzy set and $d \in [0,1]$ a real bounded variable. In fuzzification operation variable $x \in X$, the real variable d is the degree of membership in the fuzzy set s. If X is covered by N *number of* fuzzy sets, like $s_1, s_2, ...,$ and s_N which have degrees $d_1 = f_1(x)$, $d_2 = f_2(x)$, ..., and $d_N = f_N(x)$, respectively. The fuzzification operator F maps an element $x \in X$ to the set of fuzzy variables $\{(s_1, f_1(x)), (s_2, f_2(x)), ..., (s_N, f_N(x))\}$. For example, if the crisp values of variable X are fuzzified as shown in Figure 7.4, then the value of function F can be determined by intersection operation as

F: $-20 \rightarrow \{$(NB, 1), (NS,0), (Z, 0), (PS, 0), (PB, 0)$\}$, as there is no intersection

F: $2 \rightarrow \{$(NB,0), (NS, 0), (Z, 0.52), (PS, 0.4), (PB, 0)$\}$, as Z intersects with PS, therefore the minimum value of intersection is 0.4

The fuzzification $F(x)$ of an arbitrary element $x \in X$ can be represented by an N vector called fuzzy vectors in several equivalent ways

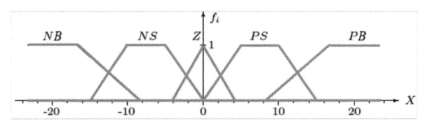

NL = negative big, NS = negative small, Z = zero, PS = positive small, PL = positive big.

Figure 7.4   Membership functions of FLC

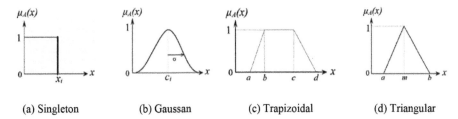

(a) Singleton (b) Gaussan (c) Trapizoidal (d) Triangular

Figure 7.5   *Types of fuzzifiers or membership functions, (a) singleton,*
(b) Gaussian, (c) trapezoidal, (d) triangular

$$F(x) = \begin{bmatrix} v_1 \\ v_2 \\ : \\ : \\ v_N \end{bmatrix} (x) = \begin{bmatrix} (s_1, f_1(x)) \\ (s_2, f_2(x)) \\ : \\ : \\ (s_N, f_N(x)) \end{bmatrix} \cong \begin{bmatrix} f_1(x) \\ f_2(x) \\ : \\ : \\ f_N(x) \end{bmatrix} \tag{7.4}$$

In the last vector, the fuzzy sets s_i are not explicitly noted down but are by the indices i.

The fuzzifier in FLC converts crisp value of variables to fuzzy values. There are largely four types of fuzzifiers (Figure 7.5) used to control SMES.

(a) A singleton fuzzifier maps an object to the singleton fuzzy set centred at the object itself (i.e., with support and core being the set containing only the given object) as shown in Figure 7.5a. This fuzzifier is used if any specific crisp value to be focused at the position x_i.

(b) A Gaussian fuzzifier always has a maximum value of 1 as shown in Figure 7.5b. A Gaussian fuzzifier can be represented as

$$\mu_A(x) = exp\left(-\frac{(c_i - x)^2}{2\sigma^2}\right) \tag{7.5}$$

where c_i and σ are the centre and width of the fuzzy set, A, respectively. Gaussian fuzzifier is applied when concise and smooth notations are required. Gaussian fuzzifier has the advantage of being non-zero at all points.

(c) Trapezoidal function as shown in Figure 7.5c is defined by a lower limit a, an upper limit d, a lower support limit b and an upper support limit c, where $a < b < c < d$.

$$\mu_A(x) = \begin{cases} 0, & (x < a) \ or \ (x > b) \\ \dfrac{x-a}{b-a}, & (a \le x \le b) \\ 1, & (b \le x \le c) \\ \dfrac{d-x}{d-c}, & (c \le x \le d) \end{cases} \tag{7.6}$$

Table 7.2   Fuzzy set operations

Operations	Equations	Figures
Union	$\mu_{A\cap B}(x) = max(\mu_A(x), \mu_B(x))$	
Intersection	$\mu_{A\cup B}(x) = min(\mu_A(x), \mu_B(x))$	
Complement	$\mu_{\bar{A}}(x) = 1 - \mu_A(x)$	
Cartesian product	$\mu_{A_1 \times \cdots \times An}(u_1, u_2, \cdots u_n) = min\{\mu_{A_1}(u_1), \cdots, \mu_{A_n}(u_n)\}$ or $\mu_{A_1 \times \cdots \times An}(u_1, u_2, \cdots u_n) = \mu_{A_1}(u_1), \mu_{A_2}(u_2) \cdots, \mu_{A_n}(u_n)$	
Fuzzy relation	$R_{U_1 \times \cdots \times U_n} = \{((u_1, \cdots, u_n), \mu_R(u_1, \cdots, u_n)) \mid (u_1, \cdots, u_n) \in U_1 \times \cdots \times U_n\}$	
Sup-star	$R \circ S = \{[(u, w), sup(\mu_R(u, v) * \mu_s(v, w))], u \in U, v \in V, w \in W\}$	

There are two special cases of a trapezoidal function, which are called R-functions and L-functions:

R-functions: with parameters $a = b = -\infty$

L-functions: with parameters $c = d = +\infty$

(d) Triangular fuzzifier is defined by a lower limit a, an upper limit b and a value m, where $a < m < b$ as shown in Figure 7.5d.

$$\mu_A(x) = \begin{cases} 0, & x \le a \\ \dfrac{x-a}{m-a}, & a < x \le m \\ \dfrac{b-x}{b-m}, & m < x < b \\ 1, & x \ge b \end{cases} \tag{7.7}$$

In fuzzy logic control, fuzzy rules are used to define the map from the fuzzified input signals (error signals, measured signals or command signals) of the fuzzy controller to its fuzzy output signals (control signals). In a single-input single-output (SISO) fuzzy rule, an input error signal E determines the fuzzy set $s_e = (E, f_e)$ and a control signal determines a fuzzy set $s_u = (U, f_u)$. The SISO rule mapping the fuzzy input variable $v_e = (s_e, d_e)$ to the fuzzy output variable $v_u = (s_u, d_u)$ of the fuzzy controller is defined as $r_u = (s_u, d_e)$.

The fuzzy rules that are applied for fuzzy-controlled output are AND rule, OR rule or different complicated fuzzy rule as shown in Table 7.2. Two input error signals, e_1 and e_2 contain two fuzzy sets $s_{e1} = (E_1; f_{e1})$ and $s_{e2} = (E_2; f_{e2})$, and the control signal u contains a fuzzy set of $s_u = (U, f_u)$. The rule mapping the fuzzy input variables $v_{e1} = (s_{e1}, d_{e1})$ and $v_{e2} = (s_{e2}, d_{e2})$ to the fuzzy output variable $v_u = (s_u, d_u)$ is defined as $r_u = (s_u, min (d_{e1}, d_{e2}))$. $s_{e1} \cap s_{e2} \rightarrow s_u$, where the degree of firing is $d_u = min (d_{e1}, d_{e2})$. Fuzzy associative memory (FAM) is the collection of all fuzzy rules. Each of the fuzzy rules is evaluated for every control cycle through parallel processing. The output of each fuzzy rule is a fuzzy variable. The output of the FAM is equal to the vector sum of all these fuzzy variables. The table shown in Table 7.3 representing all the fuzzy rules is called FAM table.

Defuzzification of the fuzzy output variable from Fuzzy rule is essential since the system parameters cannot be controlled with linguistic variables, like 'high' or 'low' from fuzzy controller, but requires specific signal. Defuzzification is the alteration of a fuzzy linguistic variable to a precise signal, just as reverse operation of fuzzification. The fuzzy output is produced from the logical union of two or more

Table 7.3   Fuzzy associative memory

		Input1 (e₁)
Input 2 (e₂)	Fuzzy Membership (v₂)	Fuzzy Membership (v_{e1}) Fuzzy Rule consists (r_u)

Table 7.4   Methods of defuzzification

Defuzzification methods	Equations	Figures
Max membership principle	$\mu_C\left(z^*\right) \geq \mu_C\left(z\right)$, for all $z \in Z$	
Mean max membership	$z^* = \dfrac{a+b}{2}$	
Weighted average method	$z^* = \dfrac{\sum \mu_C\left(\bar{z}\right) \cdot \bar{z}}{\sum \mu_C\left(\bar{z}\right)}$	
Centroid method	$z^* = \dfrac{\int \mu_C(z) \cdot z \, dz}{\sum \mu_C(z) \, dz}$	

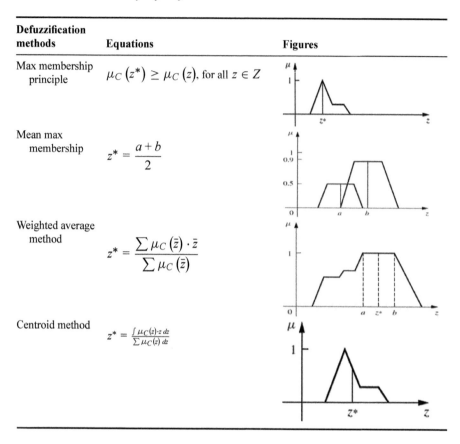

fuzzy MFs defined on the universe of discourse of the output variable. The methods of various defuzzification methods are given in Table 7.4.

There are two types of fuzzy interface systems mainly used for controlling SMES for power system applications. These are Mamdani and Tgaki–Sugeno fuzzy interface systems as shown in Figure 7.6. These methods are the mathematical procedures to perform logical interfacing of the IF-THEN rules. Mamdani model is used for both linear and non-linear systems whereas the Tgaki–Sugeno is used for non-linear system only.

In Mamdani method, the inputs of crisp values are considered in rule-based logic. The rules interface with the outputs that are also crisp values. If the i_1 and i_2 are the inputs of FLC and o is the output, then these are related by r linguistic proposition of Mamdani form given by

IF i_1 is

$$A_1 \text{ and } i_2 \text{ is } A_2 \text{ THEN } o \text{ is } B, \text{ for } K = 1, 2, \dots, r \tag{7.8}$$

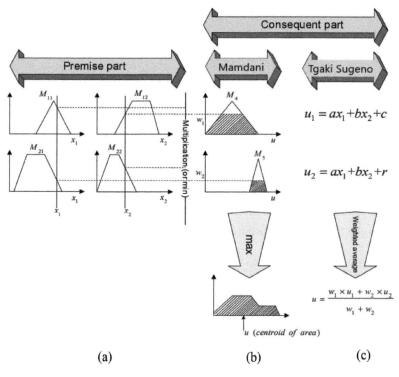

(a)   (b)   (c)

Figure 7.6   *Methods of fuzzy interface system: (a) premise part, (b) Mamdani,*
                (b) Tgaki–Sugeno

where A_1 and A_2 are the fuzzy sets representing the kth inputs, and B is the fuzzy set representing kth output. The MFs can be represented by

$$\mu(i_1) = T(i_1 - input(x)) = \begin{cases} 1, i_1 = input(x) \\ 0, otherwise \end{cases} \tag{7.9}$$

$$\mu(i_2) = T(i_2 - input(y)) = \begin{cases} 1, i_2 = input(y); \\ 0, otherwise. \end{cases} \tag{7.10}$$

T represents the triangular MF; *input* (x) and *input* (y) are the input variables. The output based on the Mamdani implication can be given as

$$\mu_B(y) = max\left[min\left[\mu_{A_1}(onput(x)), \mu_{A_2}(onput(y))\right]\right], \ k = 1, 2, ..., r \tag{7.11}$$

The centre of gravity (COG) type defuzzification is mostly used for Mamdani-type fuzzy interface.

If Tgaki–Sugeno is applied for the same inputs i_1 and i_2 then the rule can be expressed as

$$IF \ i_1 \ is \ A_1 \ and \ i_2 \ is \ A_2, \ o = f(i_1, i_2) \tag{7.12}$$

where o is the crisp output function within the region mentioned in fuzzy rule. While designing the fuzzy model using input output data, it requires structure identification and parameter identification that are very complicated for non-linear system. To overcome this, model can be derived from the equation of given non-linear system.

Though the power system is a very complicated non-linear system, the equations that represent SMES during handling the issues of the power system are linear. Therefore, Mamdani fuzzy interface is applicable for SMES. In case of Takagi–Sugeno, the first-order model can also be used as it gives linear output. The output of Sugeno interface is a function obtained by weighted average defuzzification. This consumes less time than for Mamdani fuzzy interface as it directly gives crisp output in linguistic form, whereas in case of Mamdani fuzzy interface, output originates in a fuzzy form.

7.4   Neural network-based control of SMES for power grids

The idea of artificial neural networks (ANN) is taken from the structure and performance of the biological neural networks (BNN) as shown in Figure 7.7 a. It consists of three important active parts: the receptors, the NN and brain, and the effectors. Information from the outside world or from the internal systems passes through the receptors to the NN and brain in a form of electrical impulse. The brain makes the proper decision and passes to the effectors. The effectors transform the impulse from

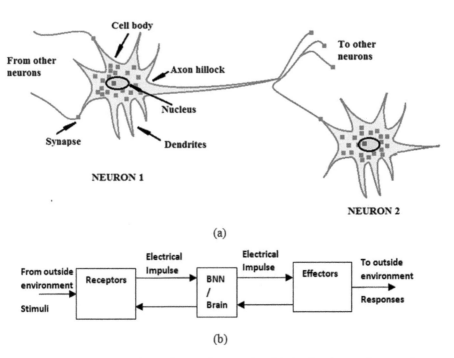

Figure 7.7   (a) Biological neural networks; (b) Artificial neural networks

the brain into response to the outside environment. The primary part of an NN is the neuron. It consists of soma, axon and dendrites. The attractive features of BNN that make it superior than any advanced artificial intelligent system are robustness, tolerance of fault, speed adaptability, flexibility that can deal with inconsistent, fuzzy and noisy data and can do parallel operation. Figure 7.7b is an assemble of processing units of interconnected network, which shows similar features like BNN.

ANN is generally used for solving classification problems (identify images, recognition of face expression, voice detection, etc.) and clustering problems (comparing items, anomaly detection, etc.). In clustering algorithm, the input data are interpreted to find the featuring group. The classification problem uses data set based on human knowledge and trains ANN to make a correlation between the given data and target. This uses supervised learning. For using ANN as a controller for SMES, supervised learning is required. The learning process of an ANN is also called optimization algorithm. The performance of an ANN depends on network architecture, learning algorithm, training data, number of features etc.

ANNCs are mainly intended to follow the learning activities of neural systems of the human brain. Therefore, a properly trained ANNC can adeptly take in hand non-linear systems, like electrical power systems. The steps of using ANNC are choosing the network architecture, selection of architecture specifics and training NN. The architecture of an ANNC is chosen based on the system (steady state or dynamic) to be controlled. For problems of dynamic modelling and control, ANNC network is popular. It has the ability to act as a predictor of following input signals in addition to a non-linear filter of noise. Therefore, it can be used as a controller of non-linear dynamic systems.

ANNC is used in various power system applications such as load forecasting, wind power/speed forecasting, photovoltaic electricity generation forecasting, locating and diagnosing faults, monitoring voltage stability, power system stabilization, load flow studies, power quality disturbance and state estimation [8–10].

Architecture of NN for SMES is a mathematical representation of biological neuron. NN is also referred as perceptron. Single-layer perceptron and multilayer perceptron (MLP) are the two basic types of NN architecture that are commonly used. To avoid the complexity of multilayer NN, single-neuron network can be used for controlling SMES [9]. Then multiple numbers of NN with single neuron are required to be designed. Each single-neuron NN is to control the deviation in each variable. For example, a single-neuron NN can be designed to sense the speed deviation for controlling the active power injection of SMES, whereas another single neuron can be designed to sense the voltage regulation for controlling reactive power injection of SMES. This control is designed mainly for the improvement of transient stability and voltage regulation. If ANNC is designed and trained with the results of classical controller for converter circuit of SMES then also the prior gives better result for SMES than the later one [10]. Multilayer ANNC can be designed to control both the active and reactive powers of SMES according to the rotor angular velocity deviation and the voltage deviation of generator, respectively. Types of network architectures used in power system applications are

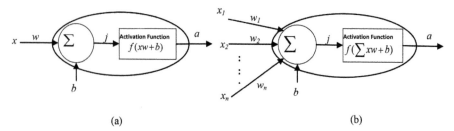

Figure 7.8   *Single neuron (a) with single input, (b) with multiple inputs*

1. single node with its own feedback,
2. single-layer feedforward network,
3. single-layer recurrent network,
4. multilayer feedforward network, and
5. multilayer recurrent network.

Among these, multilayer feedforward network (MLFFN) or MLP is mostly used as it gives better result compared to the others in power applications. The MLFFN structure consists of an input layer, multiple hidden layers (HL) and an output layer. An NN for SMES mainly has three types of layers: (1) the input layers that identify error in power system variables as an input; (2) the HL take input from another layer. The neurons are updated according to the target through learning process and associated activation functions. The output from HL goes to succeeding layers; (3) the output layers gives a required control signal for SMES.

Input layer usually consists of a single layer and accepts input features, the variables from SMES and power system to be controlled. The input layer only passes the input features or information to the succeeding HL. The neurons in input layer do not contain any activation function and not participate in the learning process. The input layer consists of one or many neurons based on the number of features considered to solve a power system problem with SMES. A neuron can have one or multiple inputs as shown in Figure 7.8. A scalar input x that is having a scalar weight w, multiplies to produce the form x_w. Another input is multiplied by a bias (or offset) b. These two products are then passed to the summer. The summer output j, often referred to as the net input, goes into a transfer function (activation function), which produces the scalar neuron output a. The weight, w and bias, b are both adaptable scalar elements of the neuron. Generally, the activation function is selected by the designer and then the elements w and b will be adjusted by the learning rule so as to achieve the neuron input/output relationship to a specific target.

Same activation function is normally used for all HL. Activation functions are also typically differentiable, meaning the first-order derivative can be calculated for a given input value. There are many different types of activation functions used in NN, although only some standard functions are used in practice for hidden and output layers. Table 7.5 shows the different activation functions usually used in different layers in NN architectures. An activation function of an NN modifies the

Table 7.5   *Activation functions for neural network*

Name of activation function	Equation	Graphical representation	Layer
Log-sigmoid	$a = \dfrac{1}{1 + e^{-n}}$		Hidden
Hyperbolic tangent sigmoid	$a = \dfrac{e^n - e^{-n}}{e^n + e^{-n}}$	$a = tansig(n)$	Hidden
Linear	$a = n$		Output
Gaussian	$a = e^{n^2/\sigma^2}$		Hidden

weighted sum of the inputs and transforms into an output from a node or nodes in a layer of the network. The competency and performance of NN depend on the proper selection of activation function. In different layers of an ANNC model different activation functions can be used. In a layer, same activation function is usually used for all neurons. ANNC controller for SMES is shown in Figure 7.9.

ANNC gathers experience from the historical data or behaviour of a system that successfully managed to overcome a problem or situation. Therefore, collection of data is the most important part of ANNC modelling. The following step of architecture selection and data accumulation is the training of NN with these historical data. After successful training, an NN becomes ANNC. The training process of NN includes the initialization of the weight vectors, training the algorithm, tracking the performance index and stop training at the predefined stopping criterion. NN are typically trained using the backpropagation of error in order to update the weights of the model. This is required for the training algorithm to minimize the derivative of prediction error.

Power system is a non-linear system and suffers from very complicated disturbances. SMES needs very robust and adaptive controller to maintain the stability of power system. Single-neuron perceptron may not be sufficient for SMES to solve

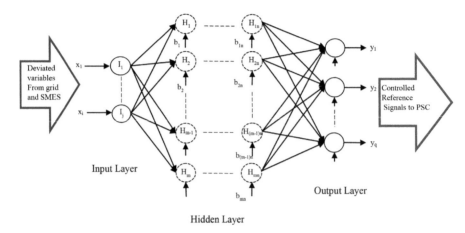

Figure 7.9   Artificial neural network for SMES

all types of issues in power system. Therefore, MLFFN is generally implemented to control SMES operations. MLFFN utilizes backpropagation training methods for learning in which information is fed forward the error for the layers are calculated and backpropagated to the previous layers. The generalization of network, depends upon the ML architecture, means that the network can work for new data, which it had not seen or trained before. For power system applications, generalization of network is very important to cope up with numerous uncertainties. There are many types of MLFFN architectures, like typical MLFFN, fully connected network (FCN) and fully connected cascade network (FCCN) as shown in Figure 7.10.

The training of MLFFN is more difficult than single neuron. In typical MLFFN, the signals pass more neurons than in other network topologies. Also, it is much easier to write algorithm for simple MLFFN than for randomly connected NN. Research is going on to introduce faster and more effective algorithms to MLFFN with more optimum results and less convergence time. The major training algorithms used to train MLFFN are

1. Backpropagation
2. Levenberg–Marquardt Algorithm (LMA)
3. Neuron by neuron or modified LMA

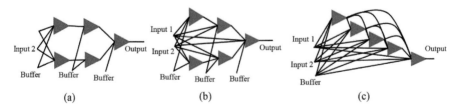

Figure 7.10   Multilayer neural network: (a) typical MLFFN, (b) FCN, (c) FCCN

In backpropagation MLFFN, the output from any neuron can be written as

$$o_p = F\{f(w_1x_{p1} + w_2x_{p2} + \cdots + w_nx_{pn})\} \tag{7.13}$$

where F is the activation function, $w_1, \ldots, w_n$ are the weights in pattern p, $x_1, \ldots, x_2$ are the inputs from the former neurons. The error from the following neuron can be represented as

$$E = \sum_{p=1}^{pn} [d_p - o_p]^2 \tag{7.14}$$

where d_p is the desired or target output. The error gradient of all the weights in pattern p is

$$\frac{d(E)}{dw_i} = -2\sum_{p=1}^{pn}\sum_{i=1}^{i} [(d_p - o_p) F'(z_p) f(net_p) x_{pi}] \tag{7.15}$$

where $F'(z_p) f(net_p) x_{pi}$ is the derivative function of $(d_p - o_p)$ and $f(net_p) = \frac{1}{2}(1 - o_p^2)$. Weight update for one single pattern p is

$$\Delta w_p = \alpha (d_p - o_p) F'(z_p) f(net_p) x_p \tag{7.16}$$

where α is the learning rate and is a non-negative factor. Weight update for all patterns applied is

$$\Delta w_p \approx \alpha \sum_{p=1}^{np} (\Delta w_p) \tag{7.17}$$

The disadvantage with the back propagation (BP) algorithm is that it has slow convergence and its inability to find optimum solutions. Also BP algorithm has difficulty to train neurons with the maximally wrong answer. Therefore, the backpropagation of the kth network can be updated as

$$w_{k+1} = w_k - \alpha g_k \tag{7.18}$$

To overcome the shortfalls of BP algorithm, ANNC can be trained using the LMA. It is the fastest algorithm to update weights and can be derived from BP and Newton algorithm as

$$w_{k+1} = w_k - (H_k + \mu I)^{-1} g_k \tag{7.19}$$

where w_{k+1} is the following weight vector value, w_k is the present weight vector value. H_k is the Hessian matrix and can be represented by Jacobeans

$$H_k = 2J_k^T J_k \tag{7.20}$$

J_k is the Jacobian matrix of $e(w)$ with respect to weight vector of NN, e is the output error of ANNC at time step n (=1 to N), which can be represented as

$$
J_k =
\begin{bmatrix}
\dfrac{\partial e_{11}}{\partial w_1} & \dfrac{\partial e_{11}}{\partial w_2} & \cdots & \dfrac{\partial e_1}{\partial w_1} \\[2mm]
\dfrac{\partial e_{21}}{\partial w_1} & \dfrac{\partial e_{21}}{\partial w_2} & \cdots & \dfrac{\partial e_{21}}{\partial w_n} \\[2mm]
\vdots & \vdots & & \vdots \\[2mm]
\dfrac{\partial e_{M1}}{\partial w_1} & \dfrac{\partial e_{M1}}{\partial w_2} & \cdots & \dfrac{\partial e_{M1}}{\partial w_N} \\[2mm]
\vdots & \vdots & & \vdots \\[2mm]
\dfrac{\partial e_{1p}}{\partial w_1} & \dfrac{\partial e_{1p}}{\partial w_2} & \cdots & \dfrac{\partial e_{1p}}{\partial w_N} \\[2mm]
\dfrac{\partial e_{2p}}{\partial w_1} & \dfrac{\partial e_{2p}}{\partial w_2} & \cdots & \dfrac{\partial e_{2p}}{\partial w_N} \\[2mm]
\vdots & \vdots & & \vdots \\[2mm]
\dfrac{\partial e_{Mp}}{\partial w_1} & \dfrac{\partial e_{Mp}}{\partial w_2} & \cdots & \dfrac{\partial e_{Mp}}{\partial w_N}
\end{bmatrix}
\tag{7.21}
$$

where I is identity matrix and μ is learning rate of LMA and can be represented as $\mu = \frac{1}{\alpha}$. When the scalar μ is 0, the iterations are faster, more accurate and having minimum error. When μ is large, the iterations become slow as this turn gradient descent (GD) to a small step size. Thus, the aim is to decrease μ after each successful step and thereby, performance function is always reducing during each iteration of the algorithm. Therefore, the grade of the performance index should always be decreasing. When the performance index reaches sufficiently small threshold, the training algorithm is stopped. The GD can be represented as

$$
g_k = 2J_k e_k \tag{7.22}
$$

From (7.24) LMA can be represented by

$$
w_{k+1} = w_k - \left(J_k^T J_k + \mu I \right)^{-1} J_k e_k \tag{7.23}
$$

where e_k is defined as

$$
e_k =
\begin{bmatrix}
e_{11} \\
e_{12} \\
\cdots \\
e_{1M} \\
\cdots \\
e_{1P} \\
e_{2P} \\
\cdots \\
e_{MP}
\end{bmatrix}
\tag{7.24}
$$

Non-linear least square problems are solved with LMA. This method is a combination of two iterative algorithms: the Gauss–Newton (GN) and the GD. To update the solution, LMA selects either GN or GD. After each iteration GD, the solution revises by selecting values to reduce the function value. More specifically, the sum of the squared errors, E becomes smaller by tracking the way of sharpest descent. In GN, μ is small, whereas in GD, μ is large. GN is more accurate and faster than GD in case of minor error. Therefore, LMA shifts to GN algorithm near the optimal value. Although LMA is having the capability of handling models with multiple unknown parameters and of finding optimal mark for distantly located starting point, it also has some drawbacks. LMA only works for MLFFN networks and can deal problems with only relatively small parameters. It becomes very slow for large sets of parameters as the size of Jacobian depends on number of parameters.

The NBN algorithm consists of two steps, forward and backward computations, to collect the information necessary for Jacobian matrix in (7.26) with pth neuron, nth weight and mth output.

$$J_{pmn} = \partial e_{pm} \partial w_n = \frac{\partial e_{pm}}{\partial o_{pm}} \frac{\partial o_{pm}}{\partial net_i} \frac{\partial net_i}{\partial w_n} = f_{pm} S_{pm} i y_{in} = f_{pm} \delta \qquad (7.25)$$

where f_{pm} is the derivative of error in pth neuron with mth output, S_{pmi} is the slope of activation function and y_{in} is the derivative of ith network with respect to nth weight, values of the signals on the neuron output nodes. The forward computation is the feed-forward signal propagation. The neurons in ANN connected to the inputs contribute in training and the output of these neurons propagated to the consecutive neurons. The following neurons then take part with the input signal obtained from previous neurons. After the completion of forward signal propagation, S_{pmi} and y_{in}, the following are stored. The operational sequence of the backward computation is reverse to that of the forward computation. The signal propagates starting from output neuron towards the input neuron. The signal can pass to the forward or backward neuron in any sequence during the forward computation and backward computation resulting the same result. Here, δ represents signal propagation of neurons during backpropagation. When y_{in} and δ values are known for different patterns, the Jacobian can be calculated form (30) and stored. NBN needs less computational time and less memory than LM because of having (1) the ability to handle arbitrarily connected NN; (2) forward-only computation (without backpropagation process); and (3) direct computation of quasi-Hessian matrix (no need to compute and store Jacobian matrix).

After choosing the NN architecture for a problem, it is required to identify the size of the NN, the number of neurons in each layer and the number of HL. Training convergence is much easier for larger NN but has very bad interpolation and generalization ability. Therefore, it cannot appropriately handle new patterns that were not involved during the training process. When a network is trained with a given data set, the data are required to distribute for training, testing and validation. The weights and the biases of the network are first trained with the training data set, which is 70% of given data. Next, the validation data set evaluates the trained network to tune with the hypothesis of the model and checks the best fitting point of data. Ideally, the case when the model makes the predictions with 0 error is said to

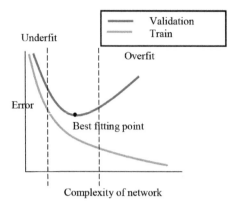

Figure 7.11   Best fitting point identification during ANNC training

have a best fit on the data. This situation is achievable at a spot between overfitting and underfitting as shown in Figure 7.11. The best fit point can be realized from the performance of our model with the passage of time, while it is learning from training data set. Regularization parameters are added in the training algorithm to achieve the best fit point. Therefore, the new objective function is developed from (7.17)

$$F = E + \lambda E_w \tag{7.26}$$

where E_w is the sum of square of the network weights; λ is the regularization parameter. The value of the regularization parameter dictates the prominence for training. Regularization parameter optimization is required for improving the generalization of ANNC. If $\lambda \gg 0$, training reduces the weight size, which may create network errors and gives a smoother response. At this time, a training algorithm of ANNC is said to have underfitting (or highly biased) when it cannot follow the essential profile of the data as shown in Figure 7.12a. Underfitting disturbs the accuracy of learning of ANNC as the model or the algorithm does not fit the given features or data well. It is important to select a model (linear or non-linear) according to the

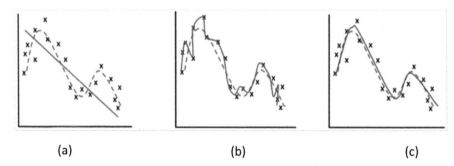

(a)   (b)   (c)

Figure 7.12   Function approximation for different values of regularization parameter λ, (a) underfit when λ is too large, (b) overfit when μ is too small, (c) best fit λ is near optimum

linearity or non-linearity of given data of features. There must be sufficient data for training the network else the learning algorithm becomes very flexible to be applied on such minimal data and therefore the model will probably make a lot of wrong predictions. Underfitting can be avoided by using more data and also reducing the features by feature selection.

Techniques to reduce underfitting (high bias)

1. increase the polynomial features thereby increasing the complexity
2. increase number of features, performing feature engineering
3. eliminate unrelated data from the data set
4. decrease the learning rate
5. increase the time of learning (i.e., number of epochs).

If $\lambda \sim 0$, then the errors from the algorithm become smaller, which may create high variance. The high variance (or overfitting) is a problem that an NN faces when the training algorithm stringently fits all the training data as shown in Figure 7.12b. It happens when the network is trained with more than the required data. The NN starts learning from the noise or inaccurate data present in the given data set. The algorithm trains the network wrongly by fitting every random data that are not related to the problem. That means the NN does not improve further during the training period and losses its ability to solve the problem. Techniques to reduce overfitting are:

1. including more training examples,
2. reducing features,
3. early stopping during the training phase shows the errors begin to increase, and
4. increasing the learning rate.

When the training algorithm best fits the data set as shown in Figure 7.12c, the model becomes more generalized, robust and can work for many new situations it is not trained with.

7.5 Adaptive neuro fuzzy inference system (ANFIS)-based control of SMES for power grids

The adaptive neuro fuzzy inference system (ANFIS) model is a concept of hybrid algorithm combining FLC and ANNC to get the advantages of both control systems. The FLC can gain self-adaptability from the low-level learning process such as backpropagation GD from ANNC with mean square error algorithm. Fuzzy logic parameters of Takagi–Sugeno fuzzy inference system are determined from training process of ANNC. Correspondingly, ANNC can increase their transparency from high-level fuzzy rule (IF-THEN) from FLC. ANFIS is developed by Jang (1993) [10].

For learning ANFIS requires training data set contains is desired input and output data of the targeted issue of the specific power system model. Figure 7.13 shows five-layer ANFIS structure. It consists of two inputs, I_1 and I_2, and one output O_p. Each input is divided into crisp value and then fuzzified into MFs. If each input has five MF, then 25 fuzzy rules are required to develop. 1in1–1in5 are five MFs of input1 and 2in1–2in5 are five MFs of input2. Initially, the shape of MFs of the ANFIS is formed, which later on can select the parameters of the functions automatically. To train these parameters, supervised learning algorithms are used based on the previous studies to handle the present power system issue. The rule base technique of these automatically selected and trained MF of ANFIS makes it more adaptive to situations. In ANFIS, each input is divided into five crisp or MFs as shown in Figure 7.14. The triangular MF of both inputs are shown in Figure 7.14a and Figure 7.14b. A first-order Sugeno fuzzy model is selected with 25 fuzzy IF-THEN rules. The output MFs are selected to give linear function. The fuzzy interface has 25 output MFs with linear functions of rth rule, $f_r = a_r x + b_r y + c_r$, where a_r, b_r and c_r are the fuzzy parameters of rth rule. These MFs are trained with hybrid training method [9] in a $2 \times 25 \times 1$ NN structure. The IF-THEN rules of fuzzy interface are

Rule 1: if I_1 is 1in1 and I_2 is 2in1 then $f_1 = a_1 I_1 + b_1 I_2 + c_1$
Rule 2: if I_1 is 1in1 and I_2 is 2in2 then $f_2 = a_2 I_1 + b_2 I_2 + c_2$

$$\vdots$$

Rule 24: if I_1 is 1in5 and I_2 is 2in4 then $f_{24} = a_{24} I_1 + b_{24} I_2 + c_{24}$
Rule 25: if I_1 is 1in5 and I_2 is 2in5 then $f_{25} = a_{25} I_1 + b_{25} I_2 + c_{25}$

Here, I_1 and I_2 are the linguistic variables. 1in1–1in5 and 2in1–2in5 are the fuzzy sets. $f_r (I_1, I_2)$ is a mathematical function. These rules are tabulated in Table 7.6. The ANNC structure of ANFISC contains one input layer, four HL and one output layer.

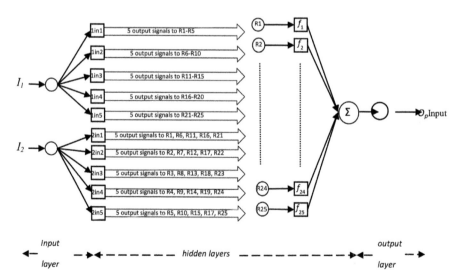

Figure 7.13   Structure of ANFIS1 with an input layer, four hidden layers and an output layer

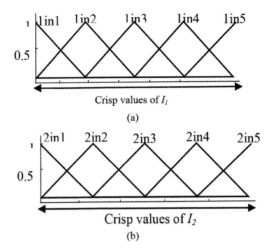

Figure 7.14   *Triangular membership functions for (a) input1 and (b) input2 of ANFIS*

The HL1 containing nodes represent the MFs of input variables. In HL1, each node is adaptive with a function:

$$O_i^1 = \mu_{1in_i}\left(I_1\right) \ for \ i = 1, 2, ..., 5 \tag{7.27}$$

$$O_j^1 = \mu_{1in_j}\left(I_2\right) \ for \ j = 1, 2, ..., 5 \tag{7.28}$$

where $1in_i$ and $2in_j$ are the linguistic labels to represent the MFs (e.g., $\mu_1 in_i(I_1)$ and $\mu_2 in_j(I_2)$) to nodes i and j, respectively. The triangular MFs are presented in I_1, for which the lowest and highest amounts are 0 and 1, respectively.

The output of this layer is fed to the HL2. In HL2, each node represents the weight of MFs. The output from each node is the multiplication of all incoming signals of the respective nodes,

$$w_{ij} = \mu_{1in_i}\left(I_1\right) \times \mu_{1in_j}\left(I_2\right) \tag{7.29}$$

HL3 is rule layer. Each node calculates the ratio of the *r*th rule's weight to the sum of all rules' weights

Table 7.6   *Membership functions of ANFIS*

I_1 I_2	2in1	2in2	2in3	2in4	2in5
1in1	f_1	f_2	f_3	f_4	f_5
1in2	f_6	f_7	f_8	f_9	f_{10}
1in3	f_{11}	f_{12}	f_{13}	f_{14}	f_{15}
1in4	f_{16}	f_{17}	f_{18}	f_{19}	f_{20}
1in5	f_{21}	f_{22}	f_{23}	f_{24}	f_{25}

$$w_{ij}^* = \frac{w_{ij}}{w_1 + \ldots\ldots + w_{25}}$$

(7.30)

In HL4, each node gives the output function

$$O_{ij}^4 = w_{ij}^* f_{ij} = w_{ij}^*(a_{ij}I_{1t} + b_{ij}I_2 + C_{ij})$$

(7.31)

where a_{ij}, b_{ij} and c_{ij} are parameter set. HL5 is a single node layer. The output functions from HL4 summarizes and computes the output signal as

$$O_1^5 = \sum w_{ij}^* f_{ij} = \frac{\sum w_{ij} f_{ij}}{\sum w_{ij}}$$

(7.32)

This layer defuzzifies the output into crisp values. The defuzzified output of ANFIS is O_p.

7.6. Case study

The integration of wind generators with the conventional power system is increasing with an aim to generate clean energy and to reduce natural fuel consumption. Therefore, it is necessary to maintain the stability and prevent the interruption of power supply to compete with the conventional system. Thus, the integrated renewable generators should have the ability to remain connected during the power interruptions during any fault. This ability of wind generators is termed as fault ride through (FRT) capability. The wind generators require additional controlling methods and external devices to adopt FRT ability. The measure of this ability is decided by the grid codes of different countries worldwide.

The controlling methods of doubly fed induction generator (DFIG) implemented mostly to rotor side converter [11–26]. These concepts are economic as the involvement of auxiliary devices can be avoided. These improved and advanced controllers of RSC are designed based on stator flux dynamics [11, 12, 14], rotor flux dynamics [18] and post fault rotor current and voltage stabilizing [21, 24, 25]. However, these control strategies sometimes show limitations in LVRT improvement for deep voltage sags due to the partial capacity of DFIG converters. Moreover, these control strategies are very complex as they need added computational efforts. To reduce this constraint of converter capacity, additional devices are required to implement with the DFIG-based WPGS system. The operations of these devices are easily understandable and the controllers are less complex.

This study is done to suggest an effective controller for SMES in a wind generator-integrated power system on the occurrence of instability. A comparative study and analysis is presented on the modern AIC control techniques of SMES to enhance the FRT capability of DFIG in same numerical computing environment. The requirement of FRT in a wind generator system arises when sudden large disturbance occurs in the power system. These disturbances can be caused by short circuit either between transmission lines or between transmission lines and ground. Due to the occurrence of short circuit, the load suddenly increases in the power system,

Table 7.7   Parameters of a DFIG and SMES

System	Parameter	Ratings
DFIG [27]	Rated power (MVA)	1.5
	Rated voltage (kV)	0.575
	Stator leakage reactance (pu)	0.023
	Rotor leakage reactance (pu)	0.18
	Mutual inductance (pu)	0.016
	RSC frequency (Hz)	1620
	GSC frequency (Hz)	2700
	DC-link capacitor (mF)	10
	DC-link voltage (kV)	1.150
	GSC and RSC controller	PI
SMES	SMES coil self-inductance (H)	10
	DC-link capacitor (mF)	0.065
	DC converter switching frequency (Hz)	100
	DC converter controller	PI
	VSC switching frequency (kHz)	1
	VSC controller	FLC, ANN, ANFIS

which wind generator cannot supply due to source constraint. Therefore, the voltage at point of common coupling (PCC) drops down. This low terminal voltage creates stress in wind generator. In this situation, it is necessary to isolate the wind genera-tor from the power system. Alternatively, an external means is applied to enhance the FRT to continue the connection during and after the fault. SMES devices, for their capability of fast response and high energy density, are recently adopted to tackle these voltage stability problems of power systems. During any fault condi-tion, SMES senses the deviation in the power system variables and sets the param-eters right by exchanging active and reactive powers.

Six 1.5 MW wind generators are integrated to the grid system through 30 km transmission line as shown in Figure 7.15. The parameters of the system are listed in Table 7.7. The system is operating with the rated wind speed of 15 m/s and delivers the rated power to the grid. The analysis of LVRT capability for the proposed sys-tem with and without SMES is done in the MATLAB/Simulink platform. The time step of solution is 50 μs and simulation time is 5 seconds. A three-phase-to-ground

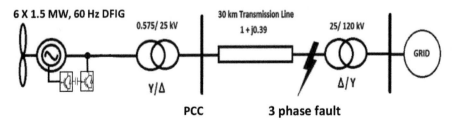

Figure 7.15   The wind generator-assorted power system under study

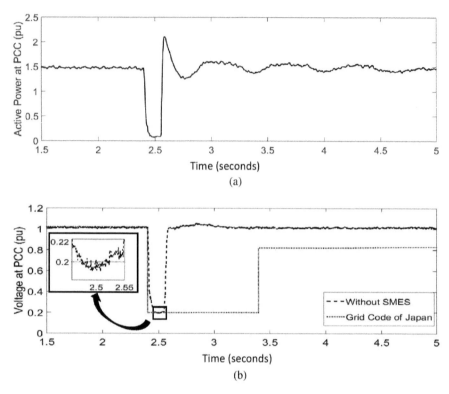

Figure 7.16   *(a) Active power at PCC; (b) voltage at PCC during three-phase fault*

fault occurred at the transmission lines at $t = 2.4$ s, as shown in Figure 7.15. After 0.15 seconds, the fault is cleared.

Due to the occurrence of fault, the power at PCC reduces abruptly as shown in Figure 7.16a from its rated value. This produces huge power demand to the DFIG wind farm. As the wind farm is operating at rated speed, it cannot supply the required power to the system. The voltage at PCC also reduces as shown in Figure 7.16b.

This voltage drop at the DFIG terminals produces large voltage disruption in the rotor circuit and surpasses the rating of rotor side converter. The current in the rotor circuit becomes uncontrollable, which produces oscillations in the electromagnetic torque. Therefore, the system fault affects and damages the rotor of the DFIG if it is not disconnected from the system. This reduces the cost-effectiveness of wind farm as it cannot produce and supply power. This isolation of wind farm reduces the active power share of the system and grid has to support the power demand during fault, as otherwise grid may suffer cascade grid fault. Recently, worldwide grid utilities have set grid codes for the voltage profile of wind farm during fault. The wind farm has to follow the voltage limit mentioned in the grid code. In Figure 7.16b, it

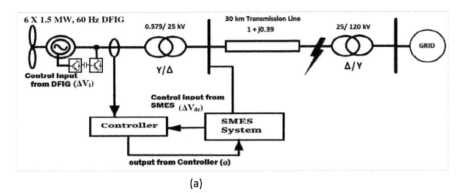

(a)

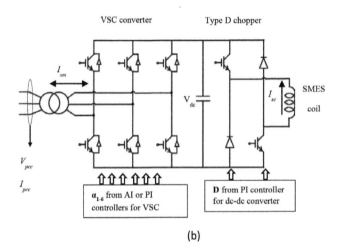

(b)

Figure 7.17   *System under study, (a) SMES connected at PCC, (b) schematic diagram of SMES*

can be seen that the voltage decreases below the grid code recommended by Japan, taken as a reference.

Figure 7.17a shows SMES, connected at the high-voltage side of the transformer. The proposed SMES unit is shown in Figure 7.17b, which includes a superconducting (SC) coil of 10 H self-inductance, a power electronic converter system consisting a B-type DC converter and a voltage source converter (VSC), a 25/0.1 kV three-phase wye-delta transformer and a 0.065 mF DC-link capacitor. The magnetically stored energy in SMES unit, E_{sm} and its rated power, P_{sm} can be expressed as

$$E_{sm} = \frac{1}{2} L_{sm} I_{sm}^2 \qquad (7.33)$$

$$P_{sm} = \frac{dE_{sm}}{dt} = L_{sm} I_{sm} \frac{dI_{sm}}{dt} = V_{sm} I_{sm} \qquad (7.34)$$

where L_{sm}, I_{sm} and V_{sm} are self-inductance, DC current and voltage of the SC coil. Values of E_{sm} and P_{sm} of proposed SMES are 500 MJ and 25 MW, respectively. The energy, E_{sm} of the coil can be delivered or stored by controlling the current, I_{sm} in the SC coil.

The SMES is controlled to operate at the time of fault and enhance the voltage as well as power of the system. The active power at PCC during fault is given in (35). The power delivered to the system by SMES is given in (7.36), (7.37). To improve the system stability, SMES is to be controlled to maintain the power equation given in (38).

$$P_{PCC} = Re\left\{V_{PCC}I_{PCC}^*\right\} \tag{7.35}$$

$$P_{sm} = Re\left\{V_{sm}I_{sm}^*\right\} \tag{7.36}$$

$$P_{sc} = V_{SC}I_{SC} = \left(L\frac{dI_{SC}}{dt}\right)I_{SC} \tag{7.37}$$

$$P_{PCC} = P_{sm} = P_{sc} \tag{7.38}$$

here P_{PCC}, V_{PCC} and I_{PCC} are the active power, voltage and current at PCC during fault. P_{sm}, V_{sm} and I_{sm} are the active power, voltage and current at the ac side of SMES. P_{SC}, V_{SC} and I_{SC} are the active power, voltage and current at the DC side of SMES. L and $\frac{dI_{SC}}{dt}$ are the self-inductance of SC coil and the rate of change of I_{sc} during fault, respectively. The required rating of SMES can be calculated from (7.33) and (7.34).

The current is controlled by varying duty ratio of the D-type DC converter between −1 and 1 according to the system requirements. To control DC converter of SMES, conventional PI controller is applied in this study. The PI controller is of D-type DC converter, shown in Figure 7.18, producing the required duty ratio (D) according to the deviation in the power output from the generator. When the duty ratio is more than 0.5, both I_{sm} and V_{sm} are positive, there by charging the SC coil. While duty ratio is less than 0.5, I_{sm} is positive and V_{sm} is negative, hence the SC coil discharges the stored energy. At persistent condition, the duty ratio is 0.5, SC coil has 0 average voltage across it and a constant persistent current.

AICs are used to control the active and reactive power exchange of VSC of SMES with DFIG integrated power system. Mamadani-type FLC with IF-THEN

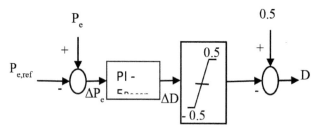

Figure 7.18   PI controller for DC–DC converter with gains K_p = 5000 *and* K_i = 0.001

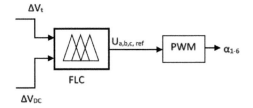

Figure 7.19   Fuzzy logic controller for SMES

rule is designed based on the behaviour and requirement of the system. The ANN and ANFIS are trained with the output of three-phase reference signals of PI controller.

FRT improvement with FLC–SMES: the FLC is an effective AIC that follows distinctive way to control the parameters of a given system. The design of FLC is easy and the control method is comprehensible. The basic structure of FLC includes fuzzification, fuzzy rules and defuzzification. Alike ANN, an FLC is implemented with two input variables and one output variable as shown in Figure 7.19. These inputs are divided into crisp values and introduced as MFs. The system complexity decides the degree of MFs. This is called fuzzification of input variables. FLC can understand only crisp or fuzzified values. The input variables (V_{dc} and V_t) and output (U_{ref}) are each having five MFs. For the input1 (V_{dc}) and output (U_{ref}), triangular MFs with overlap are used, whereas for input2 (V_t) triangular and trapezoidal MFs with overlap are used. The fuzzy sets for input1 (ΔV_{dc}), input2 (ΔV_t) and output (U_{ref}) are stated as [VL, L, ML, SL, Z], [HU, VU, U, SS, S] and [VS, S, M, B, VB] are shown in Figure 7.20(a), 7.20(b) and 7.20(c), respectively. The output is controlled by the rules defined by the user according to the system requirements. These rules then produce output in fuzzy form. Table 7.8 contains the 25 rules for the 5 × 5. The fuzzy inference system is modelled with Mamdani's max–min method and the fuzzy rule is based on the IF-THEN strategy. The output values are then defuzzified to the ana-logue values. For defuzzification process centre of the gravity method is selected.

The simulation result with FLC-controlled SMES is shown in Figure 7.21. It can be seen that the active power of PCC is improved 47% in Figure 7.21a. The volt-age at PCC also improved 63% as shown in Figure 7.21b. For these improvements SMES had to deliver 35 MW of power (Figure 7.21c) to the system and the SMES has to discharge 15 MJ (Figure 7.21d).

FRT improvement with ANNC–SMES: ANN controllers are mainly intended to follow the learning activities of neural systems of the human brain. Therefore, a properly trained ANN controller can adeptly take in hand non-linear systems, like electrical power systems. The steps of using ANN are selection of the network archi-tecture, selection of architecture specifics and training NN. The architecture of an ANN is chosen based on the system (steady state or dynamic) to be controlled. For problems of dynamic modelling and control, ANN network is popular. It has the ability to act as a predictor of following input signals in addition to a non-linear filter of noise. Therefore, it can be used as a controller of non-linear dynamic systems.

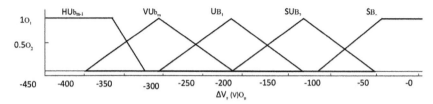

HU= Highly Unstable, VU= Very Unstable, U= Unstable, SU = Slightly Unstable, S= Stable

(a)Fig 19 Fuzzy logic controller with two input signals and

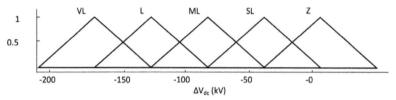

VL= Very Low, L=Low, ML= Moderately low, SL = Slightly Low, Z= Zero

(b)

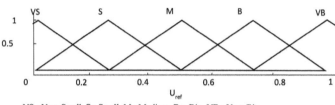

VS= Very Small, S= Small, M= Medium, B = Big, VB= Very Big

(c)

Figure 7.20   Membership functions of FLC controller (a) input1, (b) input2 and (c) output

ANN controller for SMES is shown in Figure 7.22a. To select the specifics of architecture, it is required to determine the requisite number of HL, number of neurons and size of the input vector [28]. The ANN controller has two inputs, which consist of difference between actual and reference values of generator terminal voltage (ΔV_t) and DC-link voltage (ΔV_{dc}). The ANN structure has an input layer with

Table 7.8   Membership functions of FLC

$...\Delta V_{dc}$ / ΔV_t	VL	L	ML	SL	Z
HU	M	M	VS	VS	VS
VU	M	M	S	VS	VS
U	VB	M	M	M	VS
SU	VB	VB	B	M	M
S	VB	VB	VB	M	M

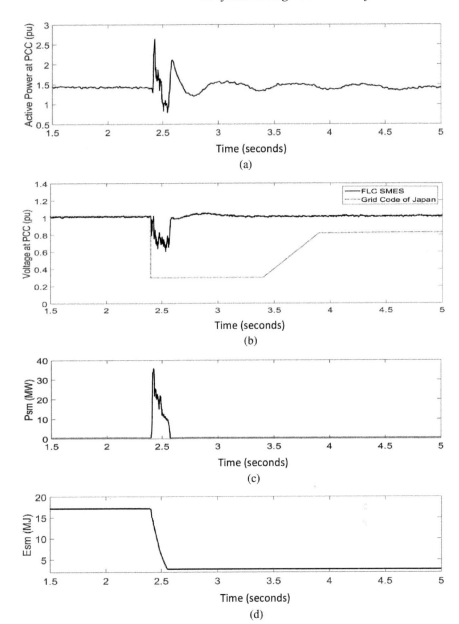

Figure 7.21   *(a) Active power at PCC with FLC-controlled SMES, (b) voltage at PCC with FLC-controlled SMES, (c) active power exchange of SMES during fault, (d) stored energy discharge of SMES during fault*

two neurons, an HL with 20 neurons and an output layer with three neurons, that is, $2 \times 20 \times 3$. This network structure is chosen by trial and error method to reach the stopping criteria. The output of each neuron depends upon the activation function. The structure of a neuron is shown in Figure 7.22b. This activation function receives inputs from $-\infty$ to $+\infty$ and sends an output between -1 and 1. The output of a single neuron [29] can be expressed as

$$a_i = f_i \left(\sum_{j=1}^{n} w_{ij} x_j + b_i \right) \tag{7.39}$$

where f_i is activation function, x_j is input vector signal, w_{ij} is weighting factor, b_i is the bias and n is the number of input vectors. Activation functions are generally applicable for both input and HL of ANN controller. Input layer processes the input variables and sends the weight to the HL. In HL, bias is added with each neuron. Tan-sigmoid function is used as an activation function of neurons in HL. The output signal of HL is transferred to the output layer through proper weighting factor. In output layer, bias is again added and linear function is used to process the

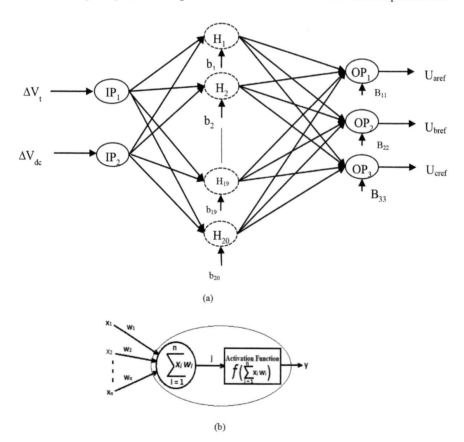

(a)

(b)

Figure 7.22 *(a) Artificial neural network for SMES; (b) single neuron*

data. Tan-sigmoid activation function is a non-linear function continuously varying between two asymptotic values (−1 and +1), which is applied to the HL and can be represented as

$$tansig\left(n\right) = \frac{2}{1 + e^{2n}} - 1 \tag{7.40}$$

Purelin linear activation function, a linear function, is used for the output layer and can be represented as

$$purelin\left(n\right) = n \tag{7.41}$$

Figure 7.23 shows the flowchart where a backpropagation algorithm is applied to train the ANNC, which updates weight and bias to get optimum output. This algorithm provides a supervised learning approach for ANNC. Training stops if any of these situations occurs: (1) the number of epochs (number of times the learning algorithm passes forward and backward of the network) reached the maximum, (2) performance is minimized to the goal, (3) performance gradient decreases below minimum limit, (4) learning rate exceeds the maximum limit or (5) validation increases after decrease in last iteration.

To use ANNC for SMES in FRT enhancement application, historical input and target data sets are required for its training. PI controller is used here to control VSC

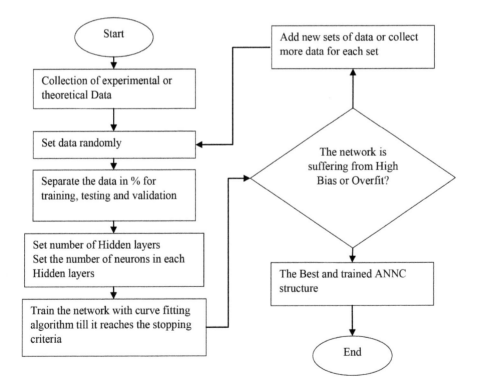

Figure 7.23   Flow chart to determine the structure of ANNC for SMES

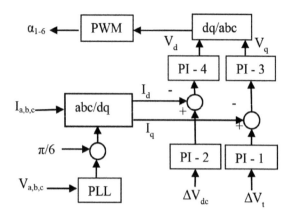

Figure 7.24   PI controller for VSC converter of SMES

of SMES to generate the data sets for ANNC. The ANNC is trained with the signals of PI controller for the SMES converter. A number of training processes are done to apply it as an SMES controller. Output layer will typically use a different activation function from the HL and is dependent upon the type of prediction required by the model.

PI controller, as shown in Figure 7.24, is employed to control the active/reactive power from VSC. The parameters of PI controller of DC converter and VSC are specified in Table 7.9. The reference signal from PI controller is converted to three-phase sinusoidal wave and compared with the triangular carrier signal to generate PWM signal for IGBT switching of VSC. The switching frequency of VSC is 4 kHz.

The simulation result with PI-controlled SMES is shown in Figure 7.25. It can be seen that the active power of PCC is improved by 27% in Figure 7.25a. The voltage at PCC is also improved by 50% as shown in Figure 7.25b. For these improvements, SMES had to deliver 30 MW of power to the system and the SMES has to discharge 10 MJ energy. There is no power exchange before and after the fault in SMES. During fault, SMES absorbs the active power at PCC. Thereby, fault current is absorbed by SMES (I_{sm}). At the same time, current is discharged from the SC coil (I_{sc}) sensing the deviation in power at PCC. These currents (I_{sm} and I_{sc}) together increase V_{dc} of VSC at SMES. This improves the voltage at PCC and the LVRT capability of DFIG. Without SMES, active power at PCC falls down rapidly during fault. Improvement in active power is seen with SMES. PI-controlled SMES improves the active power slightly. The voltage dip without SMES is lower than

Table 7.9   PI controller parameters

Controller gain	PI-11	PI-12	PI-21	PI-22
K_p	0.005	0.55	0.8	0.09
K_i	0.15	2500	500	500

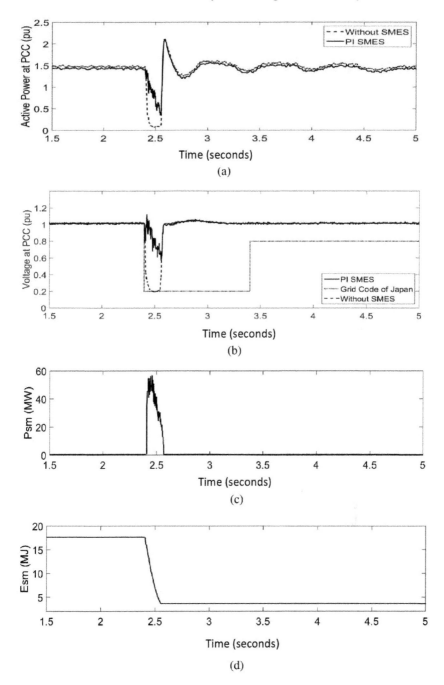

Figure 7.25 *(a) Active power at PCC with PI-controlled SMES, (b) voltage at PCC with PI-controlled SMES, (c) active power exchange of SMES during fault, (d) stored energy discharge of SMES during fault*

the minimum tolerable voltage of the Japan grid code. Therefore, DFIG cannot ride through the fault without SMES. Instead, with SMES, the voltage dips are within the acceptable limits of grid code. With AIC-controlled SMES, the improvement is better than PI-controlled SMES. With adaptive neuro fuzzy controller (ANFC)-controlled SMES, the voltage dip can be nearly eliminated.

As shown in Figure 7.25c, there is no power exchange before and after the fault. During fault, SMES is absorbing the active power. Thereby, fault current is absorbed by DC-link capacitor of SMES. During fault, SC coil discharges the stored energy (Figure 7.25d) and supplies current to mitigate the power deviation as shown in Figure 7.25a. This current along with fault current increases V_{dc} of VSC at SMES. This improves the voltage at PCC and the FRT capability of DFIG.

After designing the PI controller and simulating the wind power generator-integrated power system with PI-controlled SMES, the data input and output to and from the PI controller are collected, respectively. These data are used as training data for the ANNC multilayer feedforward network as shown in Figure 7.22a. The data from PI controller are randomly divided for training, validation and testing into 70, 15 and 15% ratio, respectively. Curve-fitting method in NN toolbox of MATLAB is used for training of the ANNC network as shown in Figure 7.26a. This ANN controller is trained using the LMA to update weights. In this study, the performance index for NN training is the mean squared error (MSE). To get the best ANN, the network is trained, tested and retrained until minimum MSE value (7.47) and highest regression, R^2 value (7.48) are reached.

$$MSE = \frac{\sum_{ts=1}^{T_s} \left(p_{ts} - t_{ts}\right)^2}{T_s} \tag{7.42}$$

$$R^2 = \frac{\sum_{ts=1}^{T_s} \left(p_{ts} - t_{ts}\right)^2}{\sum_{ts=1}^{T_s} \left(t_{ts} - t_a\right)^2} \tag{7.43}$$

where T_s denotes the number of time index t_s, p_{ts} is predicted output of the t_sth index, t_{ts} is the actual value of the t_sth index, t_a is the mean value of all outputs. When the performance index reaches sufficiently small threshold, the algorithm stops the training. The training result is shown in Figure 7.26b and 7.26c. In the linear regression, the training data are not very close but almost around the fit data. It indicates that the chance of overfitting of training data is very less. The training algorithm is stopped when the validation error starts increasing after 261 epochs as shown in Figure 7.26c. The training, validation and testing errors are $7.8e^{-3}$, $8.3e^{-3}$ and $8e^{-3}$, respectively. The regression for training, validation and testing error is $9.3e^{-1}$, $9.2e^{-1}$ and $9.2e^{-1}$, respectively.

The simulation result with ANNC-controlled SMES is shown in Figure 7.27. It can be seen that the active power of PCC is improved 33% in Figure 7.27a. The voltage at PCC also improved 75% as shown in Figure 7.27b. For these improvements, SMES had to deliver 4 MW of power to the system and the SMES has to discharge 5 MJ. This ANNC is modelled with ANN tool in MATLAB. The following step of data accumulation and ANNC architecture selection is the network training. The training process of ANN includes the initialization of the weight vectors, training the

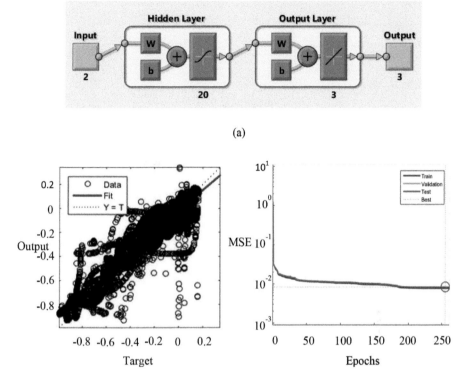

(a)

(b)

(c)

Figure 7.26   *Training of ANNC (a) neural network designed with MATLAB toolbox, (b) linear regression trained and targeted data, (c) mean square error with respect to number of epochs*

algorithm, tracking the performance index and stop training at the predefined stopping criterion. To use ANNC in LVRT enhancement application, it is trained with the signals of PI controller of SMES converter. A number of training processes are done to apply it (i.e., ANNC) as an SMES controller.

FRT improvement with ANFISC–SMES: an ANFC with two inputs and one output is designed. ANFC consists of three ANFISCs. Input signals are similar to the previous control signals. These inputs are fed to three ANFISC models simultaneously to get three-phase-controlled reference signals for PWM (as shown in Figure 7.28). However, ANFISC2 and ANFISC3 models are same as ANFISC1.

The inputs I_1 and I_2 in Figure 7.13 are replaced with the deviation in terminal voltage ΔV_t at PCC and the deviation in DC-link voltage of SMES ΔV_{dc_1} respectively. The output of each ANFISC is the controlled reference signal of PWM of SMES converter. Therefore, three ANFISC outputs give the three controlled reference signals required for the PWM converter as shown in Figure 7.28. Table 7.10 shows the crisp values of the triangular MFs shown in Figure 7.14.

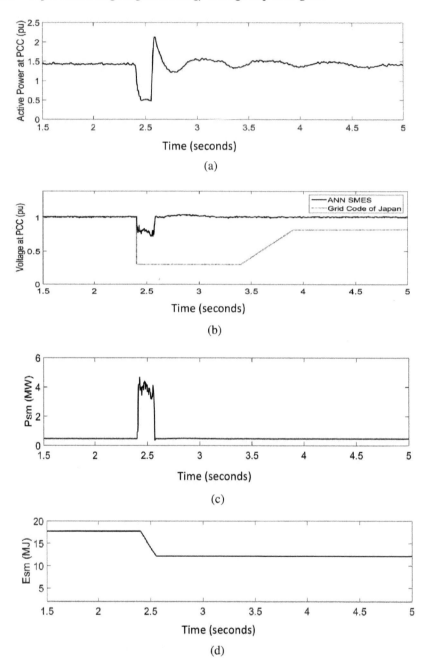

Figure 7.27   *(a) Active power at PCC with ANNC-controlled SMES, (b) voltage at PCC with ANNC-controlled SMES, (c) active power exchange of SMES during fault, (d) stored energy discharge of SMES during fault*

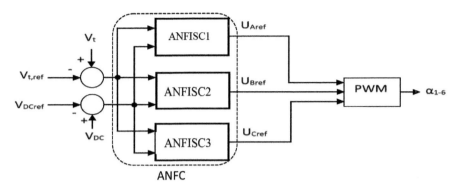

Figure 7.28   Design of ANFC with three ANFISC models

The simulation result with ANFISC controlled SMES is shown in Figure 7.29. It can be seen that the active power of PCC is improved 75% in Figure 7.29a. The voltage at PCC also improved 47% as shown in Figure 7.29b. For these improvements, SMES had to deliver 7.5 MW of power (Figure 7.29c) to the system and the SMES has to discharge 8.5 MJ (Figure 7.29d).

A comparative study, on the AIC-controlled SMES to enhance the FRT capability of DFIG-based wind generator integrated with power system, is done to suggest an effective controller for SMES with same energy storage capacity. In same numerical computing environment, the AIC controller maximizes the effectivity of SMES than conventional controller. A comparative study is shown in Table 7.11. Compared to the other FRT (or LVRT) studies, the proposed SMES with different controllers gives better results. The WPGS system under study generates more power [30,31] and more fault time [32]. From Table 7.11, it is clear that the AIC-controlled SMES provides superior control effect than conventional controllers. The ANFC-controlled SMES not only improves the active power effectively but also eliminates the voltage dip most effectively. This is due to the proper training of weight vectors and the fuzzy rules applied at the hidden neurons of ANFISC.

To enhance the LVRT capability of DFIG, the energy delivered is the maximum for fuzzy-controlled SMES, whereas ANNC-controlled SMES delivers the

Table 7.10   Fuzzy membership parameters of ANFIS for SMES controller

$I_1 = \Delta V_t$		$I_2 = \Delta V_{dc}$	
Membership	**Crisp value**	**Membership**	**Crisp value**
1in1	−800, −600, −400	2in1	−450, −300, −170
1in2	−600, −400, −250	2in2	−300, −170, −50
1in3	−400, −250, −100	2in3	−170, −50, 70
1in4	−250, −100, 0	2in4	−50, 70, 200
1in5	−100, 0, 100	2in5	70, 200, 350

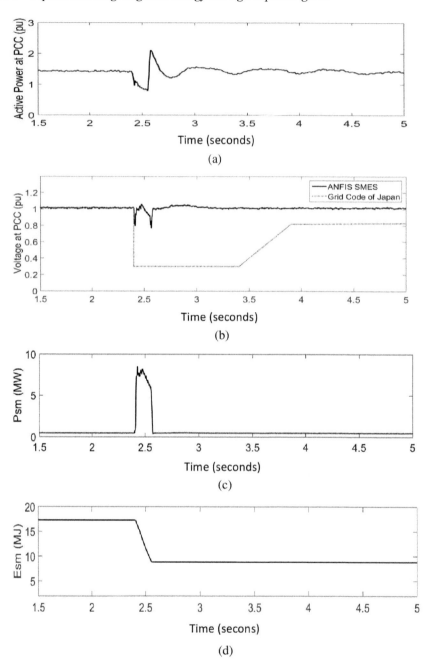

Figure 7.29   *(a) Active power at PCC with ANFIS-controlled SMES, (b) voltage at PCC with ANFIS-controlled SMES, (c) active power exchange of SMES during fault, (d) stored energy discharge of SMES during fault*

Table 7.11   Comparative stability study of SMES-controlled WPGS

WPGS rating [ref]	duration of three-phase Fault	SMES rating	SC device location	Controller	FRT improvement (%)	Energy delivered by SMES (%)
1.5 MW [30]		1.43 MJ	Inside DFIG	PI	27.3	
6 × 1.5 MW [31]	120 ms	1MJ	At PCC of wind farm	PI/FLC	38	
5 × 1.5 MW [32]	150 ms	5 × 60 kJ	DFIG conversion system	PI	10.5	
6 × 1.5 MW [proposed]	150 ms	18 MJ	At PCC of wind farm	PI/PI	50	71
				PI/ANNC	75	30
				PI/FLC	50	86
				PI/ANFC	78	50

minimum stored energy to the system. Therefore, the ANNC-controlled SMES gives cost-effective performance and can support FRT for long fault time and suitable for reliable operation of grid-connected DFIG-based wind park. In stored energy discharge and power exchange perspective, FLC is better than ANNC even though performances in voltage and power deviations are almost same. However, ANFIS requires more power and energy than FLC but less than ANN to eliminate voltage deviation for FRT. This proves that when FLC is trained to adapt the system response through ANNC, it improves the effectiveness of controller. Among different AICs, ANFC gives superior control for SMES. The AIC-controlled SMES provides superior control effect than other conventional controllers. The ANFC-controlled SMES not only overwhelms the active power effectively, but also eliminates the voltage dip entirely.

7.7 Conclusion

Power electronics converter of SMES, interfacing the coil with the AC line needs to control according to power system requirements and SC coil protection. At numerous conditions, to maintain the power quality and reliability, SMES needs to release very high power with high di/dt rate. On the other hand, reduced di/dt is preferred from the ac loss point of view to prevent the quenching of SC coil. SMES converter with low di/dt is required for voltage sag and swell, primary frequency stability and load frequency control. AICs are very effective for SMES as they imitate human thinking method. To control VSC-based SMES, AICs, such as fuzzy logic, ANN and neuro fuzzy, can give improved result, as these controllers have better handling and adaptability capability than conventional controllers.

Due to the occurrence of disturbance in power system suddenly increases the demand, the WPGS is required to be isolated as it becomes unstable and may get damaged due to sudden increase in electromechanical torque in the rotor circuit. If the WPGS is isolated, the grid becomes overloaded and unstable due to reduction of supply. The impact of the sudden occurrence of fault required to be reduced by enhancing the FRT (or low-voltage ride through) capability of WPGS. The AIC-controlled SMES at the time of occurrence of fault at the connected grid enhances the FRT capability of WPGS. A properly designed AIC can increase the fast charging and discharging capabilities of SMES to overcome the demand created during fault, thereby enhancing the dynamic stability of WPGS and increasing its connectivity with power grid. The AICs are effective, simple and easy to implement to SMES converter than conventional controllers. Moreover, the functionality of AICs is more understandable than conventional controllers.

As SMES is extensively capable of handling other power system applications, like improving rotor angle stability, voltage stability, frequency stability, load frequency control, voltage sag and swell, load fluctuation control, spinning reserve, etc., it can be suggested that AIC controllers can give better performance for all the power grid applications.

References

[1] Zhang K., Mao C., Lu J., Wang D., Chen X., Zhang J. 'Optimal control of state-of-charge of superconducting magnetic energy storage for wind power system'. *IET Renewable Power Generation.* 2014, vol. 8(1), pp. 58–66. Available from https://onlinelibrary.wiley.com/toc/17521424/8/1

[2] Khanna R., Singh G., Nagsarkar T.K. 'Power system transient stability improvement with fuzzy controlled SMES'. *IEEE International Conference on Advances in Engineering, Science and Management*; 2012.

[3] Hamdaoui H., Ramdani Y., Semmah A., Ayad A. 'Fuzzy control of ASMES to improve transient power system stability'. *IEEE Canadian Conference on Electrical and Computer Engineering.* 2003, vol. 3, pp. 1841–44.

[4] Ali M.H., Murata T., Tamura J. 'Transient stability enhancement by fuzzy logic-controlled SMES considering coordination with optimal reclosing of circuit breakers'. *IEEE Transactions on Power Systems.* 2014, vol. 23(2), pp. 631–40.

[5] Dahiya A.K., Dahiya R., Kothari D.P. 'A comparison of ANN and fuzzy for improving transient stability and frequency stabilization of SMIB'. *MIT International Journal of Electrical and Instrumentation Engineering.* 2011, vol. 1, p. 1.

[6] Xiao-hua H., Li-ye X. 'Research of fuzzy logic-controlled SMES for power system transient stability'. *2007 International Conference on Electrical Machines and Systems*; Seoul, Korea (South), 2007. Available from https://ieeexplore.ieee.org/xpl/mostRecentIssue.jsp?punumber=4411937

[7] Ambati B.B., Kanjiya P., Khadkikar V. 'A low component count series voltage compensation scheme for DFIG wts to enhance fault ride-through capability'. *IEEE Transactions on Energy Conversion.* vol. 30(1), pp. 208–17. n.d.

[8] Mukherjee P., Rao V.V. 'Power system transient stability with SMES controlled by artificial intelligent techniques'. Presented at IEEE International WIE Conference on Electrical and Computer Engineering (WIECON-ECE); Pune, India, 2016.

[9] Li Y., Cheng S., Pan Y. 'Adaptive neuron based control design for SMES unit'. *IEEE PES Transmission and Distribution Conference and Exposition.* 2003, vol. 1, pp. 217–21.

[10] Jang J.-S.R. 'ANFIS: adaptive-network-based fuzzy inference system'. *IEEE Transactions on Systems, Man, and Cybernetics.* 1993, vol. 23(3), pp. 665–85.

[11] Ardjoun S.A.E.M., Denai M., Abid M. 'A robust power control strategy to enhance LVRT capability of grid-connected DFIG-based wind energy systems'. *Wind Energy.* 2019, vol. 22(6), pp. 834–47.

[12] Eddine K.D., Mezouar A., Boumediene L., Van Den Bossche A.P.M. 'A comprehensive review of LVRT capability and sliding mode control of grid-connected wind-turbine-driven doubly fed induction generator'. *Automatika.* 2019, vol. 57(4), p. 35.

[13]   Zhu D., Zou X., Deng L., Huang Q., Zhou S., Kang Y. 'Inductance-emulating control for DFIG-based wind turbine to ride-through grid faults'. *IEEE Transactions on Power Electronics*. 2019, vol. 32(11), pp. 8514–25.

[14]   Cai G., Liu C., Yang D. 'Rotor current control of DFIG for improving fault ride – through using a novel sliding mode control approach'. *International Journal of Emerging Electric Power Systems*. 2013, vol. 14, pp. 629–40.

[15]   Chen W., Xu D., Zhu N., Chen M., Blaabjerg F. 'Control of doubly fed induction generator to ride through recurring grid faults'. *IEEE Transactions on Power Electronics*. 2015, pp. 1–1.

[16]   Chowdhury M.A., Sayem A.H.M., Shen W., Islam K.S. 'Robust active disturbance rejection controller design to improve low-voltage ride-through capability of doubly fed induction generator wind farms'. *IET Renewable Power Generation*. 2015, vol. 9(8), pp. 961–69. Available from https://onlinelibrary. wiley.com/toc/17521424/9/8

[17]   El-Naggar A., Erlich I. 'Short-circuit current reduction techniques of the doubly-fed induction generator based wind turbines for fault ride through enhancement'. *IET Renewable Power Generation*. 2015, vol. 11(7), pp. 1033– 40. Available from https://onlinelibrary.wiley.com/toc/17521424/11/7

[18]   Arribas J., Rodríguez A., Muñoz Á., Nicolás C. 'Low voltage ride-through in DFIG wind generators by controlling the rotor current without crowbars'. *Energies*. 2015, vol. 7(2), pp. 498–519.

[19]   Hu J., Wang B., Wang W., Tang H., Chi Y., Hu Q. 'Small signal dynamics of DFIG-based wind turbines during riding through symmetrical faults in weak AC grid'. *IEEE Transactions on Energy Conversion*. 2016, vol. 32(2), pp. 720–30.

[20]   Li R., Geng H., Yang G. 'Fault ride-through of renewable energy conversion systems during voltage recovery'. *Journal of Modern Power Systems and Clean Energy*. 2016, vol. 4(1), pp. 28–39.

[21]   Villegas Pico H.N., Aliprantis D.C. 'Voltage ride-through capability verification of DFIG-based wind turbines using reachability analysis'. *IEEE Transactions on Energy Conversion*. 2016, vol. 31(4), pp. 1387–98.

[22]   Vrionis T.D., Koutiva X.I., Vovos Nicholas.A. 'A genetic algorithm-based low voltage ride-through control strategy for grid connected doubly fed induction wind generators'. *IEEE Transactions on Power Systems*. 2014, vol. 29(3), pp. 1325–34.

[23]   Zhang J.H., Luan R., Shen H.R. 'Study on transient characteristic of the DFIG LVRT'. *Advanced Materials Research*. 2013, vol. 791–793, pp. 1832–36. Available from https://www.scientific.net/AMR.791-793

[24]   Zou X., Zhu D., Hu J., Zhou S., Kang Y. 'Mechanism analysis of the required rotor current and voltage for DFIG-based WTs to ride-through severe symmetrical grid faults'. *IEEE Transactions on Power Electronics*. 2018, vol. 33(9), pp. 7300–04.

[25]   Huang Q., Zou X., Zhu D., Kang Y. 'Scaled current tracking control for doubly fed induction generator to ride-through serious grid faults'. *IEEE Transactions on Power Electronics*. 2016, vol. 31(3), pp. 2150–65.

[26] Liu J.H., Chu C.C., Lin Y.Z. 'Applications of nonlinear control for fault ride-through enhancement of doubly fed induction generators'. *IEEE Journal of Emerging and Selected Topics in Power Electronics*. 2014, vol. 2(4), pp. 749–63.

[27] Available from https://www.mathworks.com/matlabcentral/fileexchange/57441-dfig-wind-turbine-model

[28] Heikin S. *Neural network a comprehensive foundation*. Pearson Education Inc; 1999.

[29] Yegnanarayana B. Artificial neural networks. New Delhi: Prentice-Hall of India Private Limited; 2005.

[30] Xiao X.-Y., Yang R.-H., Chen X.-Y., Zheng Z.-X. 'Integrated DFIG protection with a modified SMES-FCL under symmetrical and asymmetrical faults'. *IEEE Transactions on Applied Superconductivity*. 2012, vol. 28(4), pp. 1–6.

[31] Ngamroo I. 'Optimization of SMES-FCL for augmenting FRT performance and smoothing output power of grid-connected DFIG wind turbine'. *IEEE Transactions on Applied Superconductivity*. 2016, vol. 26(7), pp. 1–5.

[32] Shiddiq Yunus A.M., Abu-Siada A., Masoum M.A.S. 'Improving dynamic performance of wind energy conversion systems using fuzzy-based hysteresis current-controlled superconducting magnetic energy storage'. *IET Power Electronics*. 2012, vol. 5(8), p. 1305.

Chapter 8

Cybersecurity issues in intelligent control-based SMES systems

Mohammad Ashraf Hossain Sadi[1]

8.1 Introduction

A cyberattack is an action that undermines the security of computer systems and networks for malicious purposes. Cybersecurity of the power grid is a matter of great concern for governments, system designers, and management throughout the universe. Because of the increasing number of cyberattacks, the power grid is becoming more and more vulnerable, and the performance of the power grid is deteriorating. On the other hand, the Superconducting Magnetic Energy Storage (SMES) plays a very important role in enhancing the dynamic performance of the power grid, improving the voltage sag, voltage swell situations, and supporting the load during unbalanced situations. The voltage source converter (VSC) of the SMES plays the role of energy interface between the SMES and the power grid. The SMES also employ inverters for attaining proper voltage levels. However, due to the advent of the Internet of Things (IoT) with cloud and edge computing supports and for intelligent maintenance, the VSCs and inverters in the power grid are advancing toward the smart VSCs and smart inverters (SIs). SIs of the grid-tied power electronic interfaces provide cloud computing, condition monitoring, result visualization, remote control, and peer-to-peer (P2P) energy trading in advanced power systems. However, the advent of data injection attacks in the communication architecture can alter the measurement characteristics of power grids and have devastating consequences. Cyberattacks targeting the VSCs and inverters of the SMES impose new security and safety risks, specifically, maliciously intending to damage and disable the SMES. This chapter first discusses the cybersecurity issues in the power grid. Then how cyberattacks can take place in the SMES has been discussed. Cyberattacks in the SMES can take place into the control signal setpoints and direct current (DC) voltage signal measurement bias of the VSC. This chapter then investigated the effects of denial of service (DoS) and tampering

[1]University of Central Missouri, Warrensburg, Missouri, USA

the signal attacks on smart VSC of the SMES. The mitigation strategies of the cyberattacks on the smart VSCs have also been discussed in this chapter. All these discussions are also supported by case studies of cyberattacks in the SMES in a microgrid environment.

8.2   Cybersecurity issues in power grid

One of the major technological goals of countries around the world these days is to achieve net zero carbon emission by 2050 [1]. This will reduce the pollution, global warming effects, as well as decrease the supply of fossil fuels for the electric power systems. Large-scale integration of renewable energy sources (RES), distributed energy resources (DERs), like wind generators and photovoltaic (PV) sources, and energy storage systems (ESS), like SMES, will play a vital role in achieving this goal. The VSCs are the heart of the interface between these RES, DER, ESS, and the electric power grid. Moreover, wide area measurement systems (WAMSs) represent the real-time measurements that could be used to improve and maintain the stability of the power systems [2]. Traditional point-to-point communications are unable to meet the communication requirements, as the power grid is becoming more complex, increased, and interconnected. The WAMSs overcome the traditional measurement systems drawbacks by using modern wide area network or local area network technology [3]. Communication/Information infrastructure is the backbone of the wide area measurement and control (WAMC) system where different grid components relate to each other through this structure and are managed by the supervisory control and data acquisition (SCADA) system for supervisory actions and decisions. The vulnerabilities of the communication architecture are high and thus the communication network in the power systems can be exploited by cyber intruders to damage different layers because of poor and insufficient network security [4]. As power system is a complex cyber physical interconnected network, cyber intrusion in the cyber network will have a devastating impact on the physical network too. In recent years, data injection type cyberattacks have gained more attention due to their ability to forge the decision-making layers [5]. Most recently, the ransomware cyberattack on a colonial pipeline facility in Pelham, Alabama, forced the nation's largest gasoline pipeline to shut down [6], and the US government is putting a lot of effort into improving the nation's cybersecurity infrastructure [7]. Furthermore, cyber security of the modern power grid has become a major concern due to recent examples of general classes of data injection attacks such as SQL Slammer worm attack, LockerGoga ransomware, Triton, Crashoverride, Stuxnet, and Dragonfly [8]. The cyber space of the cyber physical network is an interconnected network of sensors, communication channels, and decision-making layers. The decision-making layers are mainly the primary and auxiliary controllers where the control decisions are performed for information flow. However, cyber vulnerabilities and contingencies in the control systems can cause unexpected risks to reliable data measurements. Bad data injections in the control systems can happen due to user errors, like unintended measurements, and topological errors or deliberate attacks, like injection of malicious data during the computation process [5]. The distorted data may trigger accidental operation of the physical devices like frequency controllers, voltage controllers, etc.

Moreover, in power systems, numerous critical equipment and field devices are used at the remote locations to effectively and efficiently collect the customer operating conditions and thus represent the WAMC technique. The field devices such as programmable logic controllers (PLCs), intelligent electronic devices, remote terminal units (RTUs), phasor measurement units, and smart meters have algorithms that can be manipulated by either customers or cyber intruders [9]. By hampering and tampering the data collection process, the utilities can get misleading feedbacks and thus can cause interruptions that may ultimately result in a blackout.

The communication network in the power systems can be exploited by intruders to damage different layers present in the grid. The RTUs forward the state estimation data to the SCADA center [10, 11]. The false data injection (FDI) either in the SCADA system state estimation scheme or data sent from the RTUs to the SCADA system can provoke negative consequences. There are possibilities of the false data not being detectable by the state estimation scheme or even the false data attack can be detectable without remedy.

There are mainly three types of attacks that can happen in the smart grid system. Packet drop attacks can be caused at various choke points in the communication path. DoS type attacks are mainly caused by disrupting, blocking, or jamming the flow of information through control and communication networks, and tampering communication data/signal type attacks not only delay communication but also contaminate the data in the communication. Furthermore, several other cyberattack situations can also take place, likesuch as intentional data injection into the power system through telemetered data [12], malicious modification of network stored data [13], coordinated data attack staged without being detected by the state estimation bad data detection algorithm [14], and undetected manipulation and the incorrect database parameters by cyberattacks [15].

The National Electric Sector Cybersecurity Organization Resource (NESCOR) defined many cybersecurity realistic events that can create a negative impact on power systems operation [16]. Later in this chapter, three NESCOR-recommended realistic events have been considered to see the effectiveness of cyberattack on the performance of ESS.

Data injection attacks in the microgrid, fundamental control systems in the power grid, and power electronics converters focused on the evaluation of the attack effects, intrusion detection technologies, and defense strategies [4, 5, 17]. FDI attack on the voltage controllers has been considered as a major issue for the power grid [18, 19]. FDI attacks in the power grid can be framed in different stages and components. The impact of the FDI attack on automatic generation control (AGC) of the synchronous generators has become a critical issue and major concern for system operators [20, 21]. Besides the data injection attacks, the effect of measurement error signals in the AGC [22] has negative impact too in the power grid.

Cyberattacks in the power electronics devices, interfaces, and controllers are emerging fields and can cause damage to the operation of the primary voltage and frequency controllers of the DERs and ESSs. Cyberattacks can take place in the converters serving in a hydropower plant [23], wind-integrated power system [24], as well as in VSC-based microgrid [25]. Moreover, FDI attacks can take place in the embedded smart voltage controllers [4, 5, 17–19], inverters of the DERs [26], DC–DC converter-based DC microgrid [17], VSCs of the PV system [27], active and reactive power (PQ) set points of power electronics inverters [28, 29], etc.

8.3 How cyberattacks happen in SMES systems?

SIs refer to consumer electronic firmware-based embedded inverter devices that can provide such functionalities as remote over-the-air updates, communication and control support, and high-speed data sampling and acquisition, along with traditional electronic inverter operations. Inverter management system (IMS) refers to the networked interconnection of SIs, which are often equipped with ubiquitous intelligence [30, 31]. The IEEE 1547 interconnection standard ensures DER communication requirements in the power grid and microgrid, which must be administered by independent system operators and distribution system operators [32, 33].

Cyber intruders employ several methods of intrusion approaches to manipulate different components in the power grid. Sensor data manipulation attack takes place in the control platform by penetrating several adversaries. Cyberattacks in the communication links can take place in the communication stage involving routers, encoders, decoders, and inside the controller. It can take place in the form of interruption of data transmission, illegitimate access to the information logs, etc. The controller cyberattacks can take place in the form of changing the reference set points in the secondary control loops and outer control loops.

The IMS with cloud support for the VSCs of the DERs and ESS like the SMES is a new field of study. The smart VSCs of SMES can provide powerful cloud computing, result visualization, greater visibility, extended scalability, condition monitoring, and remote control of individual power electronics controllers via the Internet [34]. Furthermore, the smart VSCs of the SMES can also provide decentralized optimal control, P2P energy trading with other ESSs and DERs [34]. A cyber physical network in the presence of SMES and smart VSCs is presented in Figure 8.1.

Communication protocols, such as standard RTU, Modbus, DNP3, MQTT, CoAP, XMPP, and RPL, of the SIs use lightweight protocols and are not secure because of a lack of encryption, control, and authentication. SIs with their limited system knowledge and resources are often vulnerable against adversaries. Moreover, manufacturers design the embedded controllers of the SIs by using commercial off-the-shelf components, and so the vulnerabilities of these controllers can be exploited.

At present, commercially available SI controllers have limited computing capabilities, do not run on operating systems, rely on low-level hardware systems, and boot monolithic, single-purpose firmware [24, 26]. Therefore, despite compliance with the IEEE 1547 standard, cyberattacks can disrupt the normal operation of SI-based ESSs and remain a potential threat. Given this ongoing risk and underlying cybersecurity vulnerabilities, successful implementation of SIs in SMES is a

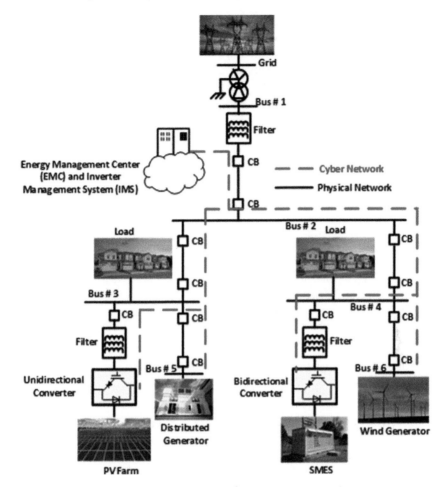

Figure 8.1   Cyber physical system with renewable energy sources, SMES, and smart VSCs

challenge for operators. Firmware attacks on firmware-controlled devices such as SMES can cause malicious behavior, may result an erroneous system output, and seek to shut down the operation of the hardware devices.

Figure 8.1 also presents the integration scheme of the grid-connected wind and PV farm along with the SMES. The wind generator is connected to the grid through the rotor side converter (RSC), grid side converter (GSC), and transformer. The PV farm is connected to the grid through a unidirectional converter interface. The SMES is connected to the bus using a bidirectional converter and VSC. The VSC of the SMES, local Energy Management Center (EMC), and IMS are connected through the cyber network. The control signals for the VSCs are processed in the EMC, and they are subject to data injection attacks. Cyberattacks on the VSC of the SMES or on the EMC can compromise the performance of power electronic converters and, hence, the overall

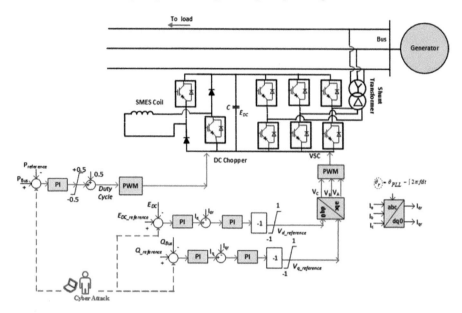

Figure 8.2   Cyberattack scenarios in the smart VSC of the SMES

grid. The possible cyberattack injection scenarios in the control system of the SMES are represented in Figure 8.2.

8.4   Impact of cyberattacks on SMES systems

Cyberattacks targeting embedded systems and PLCs have a range of socioeconomic impacts [35, 36]. Investigations into cyberattacks on power electronics-driven primary controllers of the SMES lack focus but require a proper framework and impact analysis. Data injection cyberattacks on the voltage source inverter of the SMES require detail analysis, proper focus, and impact analysis for the rest of the system as SMES plays a vital role in balancing the unbalanced situation of the power grid. Cyberattacks on the smart VSC of the SMES can be categorized as disruptive attacks, which could lead to the violation of the security boundary and instability of the cyber physical system. DoS attack refers to the disruption of data transmission, jeopardizes the operation of the devices involved, and damages the cyber physical systems' stability and security. It can cause the system to be unavailable or not function properly. For a DoS attack, the SMES may be temporarily or indefinitely locked such that its charging, discharging, load balancing, etc. operations are non-functional.

Cyberattacks in the VSC of the SMES can happen in different control signals as represented in Figure 8.2. Cyberattack in the active power (P_{BUS}) signal can increase or decrease of the active power (P_{BUS}) set points by data injections. It corresponds to intrusion injection attack into the controller, which can contravene the system objectives and represent a DoS type of attack. Increase or decrease of the reactive power (Q_{BUS}) set points by data injection attacks can also take place. It also corresponds to an intrusion injection attack into the controller, which can contravene the system objectives and represent a DoS type of attack. Adding a DC bias to the sensed DC voltage (E_{dc}) measurements corresponds to a sensor measurement attack on a DC voltage sensor via data acquisition and represents a tampering the signal type of attack.

There are three operating modes of the SMES: (1) standby mode: when the voltage across the SMES coil is equal to zero, current through the SMES coil is held constant at rated value, there will be no energy transfer between the SMES and the AC grid, and maximum energy will be stored within the superconducting coil; (2) discharging mode: when the voltage sag disturbance occurs at the grid side, SMES coil current decreases, and energy stored across the SMES coil will be delivered to the AC grid; (3) charging mode: when the voltage swell disturbance occurs at the grid side, SMES coil current increases, and energy is transferred from AC system to the SMES until maximum coil energy capacity is reached.

Cyberattacks in the primary controllers of the SMES during the standby mode can manipulate the controller setting values, which can inadvertently discharge the energy from the SMES coil, increase/decrease the voltage across the SMES coil, and disturb the steady nature of the current through the SMES col. This may unintentionally initiate the operation of the protective devices in the microgrid, initiate the islanded operation of the microgrid, and halt the operation of the DERs from feeding the loads. Persistent cyberattacks in the SMES controller during the discharging mode may lead to interrupted energy discharge operation, to enough discharge of energy from to SMES to the AC system to support the voltage sag, and even cause further severe voltage sag situation in the AC system by initiating the energy consumption operation of the SMES. Finally, cyberattacks during the charging mode of the SMES may not allow the SMES coil to charge to its maximum coil capacity, interrupted charging operation of the SMES coil, and even initiate the discharge operation instead of the charge operation of the SMES.

8.5   Detection of cyberattacks in SMES system

Cyberattacks in the primary power electronic controllers of the SMES can be detected using various detection methods and they can be broadly categorized in model-based detection, data-driven detection, and network- and firmware-based detection.

8.5.1   Model-based cyberattack detection methods

The model-based detection methodologies are more mature and implementable as they have been implemented extensively for cyberattack detections in the industry. These methods have less computational burden and relatively accurate during the cyberattack detection preparation and real-time implementation stages. The model-based detection methods are established based on the system configuration, system parameters, and model structures, which are easy to obtain and compute. Therefore, if the detection method based on the system model is built correctly, it can trigger the cyberattack detection performances. Thus, the requirement of real system data is minimal, which are sometimes difficult to obtain. The model-based detection methodology develops a physics-based model that exhibits the system in the presence of SMES under steady state condition to compare it with the actual system measurements during anomalies by determining the inconsistencies between the modeled operation and the actual operation [37]. The anomalies are determined by comparing the actual performance against a predefined residual threshold.

If cyberattack takes place in the smart power electronic interfaces of an SMES and the compromised input and output signals are w_a and y_a, respectively, then the residual norm, r, can be defined as [38]

$$|r| = y_a - h\left(\hat{x}, w_a\right) \tag{8.1}$$

where x is the system state based on the based on the output measurements y_a. During the cyberattack, the residual norm $|r|$ will produce larger than the normal value. So, a residual threshold is defined in the beginning of the algorithm to check the presence of any anomaly in the system. The algorithm can detect any cyberattack when the residual norm $|r|$ becomes more than the predefined threshold, otherwise the system is considered to operate normally.

Similarly, as part of the model-based detections, dynamic estimation methods potentially can be utilized to detect any cyberattacks in SMES interfaces as it has unique dynamics arising from the SIs under different control operations. Any cyberattacks in the SIs of the SMES will deviate its dynamic behavior from normal operating conditions. In dynamic estimation method, the estimated system model output y_a is compared with the actual received output y_a to determine the presence of any cyberattacks [39]. A Euclidean detector can determine the discrepancies between the measured signal and estimated signal and trigger any alarm if the difference is significant enough than the normal situation [39]. The significant enough difference indicates the statistical characteristic of the obtained residual.

The FDI methods are also part of model-based cyberattack detection methodologies. The FDI methods usually construct a state observer and parity equations to produce a residual for detecting cyberattacks [40–42]. The constructed parity equations conduct a detectability analysis to determine the subset of the system equations and determine whether it contains enough redundancy to generate residual for detecting the cyberattacks.

The hypothesis testing-based cyberattack detection methods can also compute the probability of the cyberattacks existence based on several hypothesis conditions. The hypothesis testing-based methods use robust optimization tools to describe the probability of a cyberattack, which can be incited by a system adversary [43, 44].

8.5.2   Data-driven cyberattack detection methods

In contrast with the model-based cyberattack detection methods, in data-driven cyberattack detection methods, detailed model information is not required. Thus, the detection performance of the data-driven methods is not affected by the model accuracy. Also, most of the data-driven detection models can be trained offline and implemented online. Although historical data are required to train the detection models, real-world data can be utilized to perfectly train the detection methods. The data-driven detection methods have been widely implemented and proposed for anomaly detection in the power grid, but their applications for the device-level and system-level controllers of the SIs require more attention.

There are different data-driven anomaly detection methods available including reinforcement learning method [45], stacked auto encoder [46], deep neural networks [47], convolutional neural network [48], vector autoregressive method [49], dynamic Bayesian network [50], support vector machine [51], principal component analysis (PCA) reconstruction [52], etc. These data-driven anomaly detection methods can identify the anomaly in the system based on the monitoring data and can be implemented for detecting cyberattacks in the smart power electronic interfaces of the SMES. The supervised learning algorithms with heuristic feature selection [53] can also be applied to detect the cyberattacks in the SIs; however, supervised learning methods require large amount of data for training the algorithm. On the other hand, the unsupervised learning algorithms can cluster the data in different classes according to specific features and they are getting popular recently. The statistical data-driven cyberattack detection methods [27, 54] can also detect and diagnose the cyberattacks in the SMES power electronics controllers. The statistical data-driven methods can detect the cyberattack based on the constructed and analyzed high-dimensional electric waveform data feature matrix.

Besides the cyberattack detection, the data-driven methods can also distinguish between the normal operation and faulty operation of the SMES, data integrity attacks (DIA) etc.

8.5.3   *Network- and firmware-based cyberattack detection methods*

According to the IEEE Std. 1547–2018, DERs must use Modbus, Distributed Network Protocol 3 (DNP3), or Smart Energy Profile 2.0 as their communication protocol [55]. However, the network-, firmware-, and software-related vulnerabilities can bypass the most sophisticated firewall, access control, and security defense mechanisms [56]. But to the best of our knowledge, so far these firmware-related cybersecurity issues have not been extensively explored specially for the SIs as compared with the analysis reported for the network-related cybersecurity of the power grid. As the SIs coordinate their operations using the communication networks, the established network- and firmware-based cyberattack detection methods for the SIs can as well be applied for the SMES. The Sandia National Laboratory investigated three network-based defense strategies specifically for the inverter-enabled DERs, which are also applicable for the SMES [57]. The methods are categorized as network segmentation, encryption, and moving target defense. Furthermore, the National Renewable Energy Laboratory (NREL) also reported several reports to mitigate the network-related attacks in SI-enabled power grid such as the strong key management and user access control, public key infrastructure (PKI), etc. [58]. The NREL also proposed an open source hardware security module termed as "module for operational technology (Module-OT)" incorporated into the inverter hardware to strengthen the network security [55]. The Module-OT can be integrated in the transport layer within a communication network of the open systems interconnection model. To detect the real-time network intrusion for the inverter-enabled systems, several signature and rule-based network intrusion tools have been reported such as Snort, Suricata, etc. [59, 60]. These tools can detect irregular network packet format, reply, and authenticate the messages. Also, watermarking and perturbation-based diagnostics can also determine the real-time network intrusion for SI-enabled ESS [61, 62]. The hardware performance counter (HPC) method [63] can identify the malicious malware modification cyberattacks in SIs by periodically determining the order of different instruction types within the SI firmware codes and finally identifying any unwanted code modifications using machine learning (ML) classifiers.

The blockchain security technology has been widely adopted for the communication system security, data integrity, and software security in the IoT and e-commerce systems [64]. According to Mylrea and Gourisetti [65], the blockchain security can meet the critical infrastructure software patching requirements of the North American Electric Reliability Corporation. Thus, it can provide secure distributed framework for the inverter-enabled communication networks by providing updated cryptography, PKI, better key management, patch management, consensus, access control mechanism, etc. A private zero-trust ecosystem blockchain network for the SI-enabled power network including the SMES is represented in Figure 8.3.

In blockchain technology, the zero-trust ecosystem means that no network-related entity can be trusted including the devices, applications, data, etc. [66]. In Figure 8.3, a private blockchain network has been represented where multiparty including the utilities, operators, vendors, service providers, etc. can seamlessly collaborate with each other to maintain a secure ecosystem. The multiparty in the blockchain network can seamlessly coordinate with each other through effective notifications, disclosures, and validation mechanisms to identify the user- and vendor-related incidents while keeping the privacy of the system through smart contacts and multichannel blockchain. Security modules are located and installed in critical devices such as inverters, site data managers, cloud platforms, etc. In the security modules, there are blockchain client program, intrusion detection system (IDS), static malware analysis, and firmware patch. Thus, each critical device in the blockchain network will have its own identification (ID), asymmetric key, and smart contract-based control access resulting in less complicated key management. The smart contracts can control the access ledger, PKI, and enable the submission of transactions between different blockchain clients [67]. Finally, using the security modules, the critical devices and critical assets like the inverters can consistently verify the identity, limit the access, and validate the activities to ensure the authenticity and integrity of the cyber space of the power network.

The blockchain technology can prevent the DIA as part of critical asset protection. The blockchain keeps the security logs stored in the ledger to authenticate, authorize, and verify the in-transit data. The multiparty private clients, like the utilities, operators, vendors, and security service providers, are authorized entities to provide data as a form of transactions to the blockchain network and can share the data stored in the blockchain ledger. For the authorized clients, accessing the blockchain ledgers can create transactions that include hash values of control commands or files. These control commands and file updates are considered as authorized events in the blockchain network. After the hash values are provided to the authorized events in the blockchain ledger, the smart contracts proceed the integrity

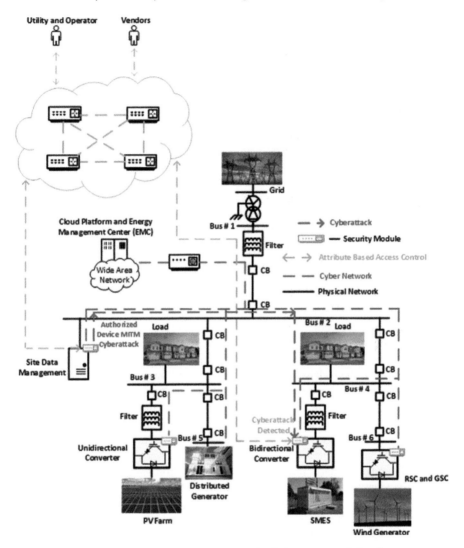

Figure 8.3 *A private zero-trust ecosystem blockchain network for the smart inverter-enabled power network*

checks. So, the overall blockchain network can provide increased visibility, services, and applications to ensure the data integrity, and authenticity of the commands.

8.6   Mitigation of adverse effects of cyberattacks on SMES

IDS is the technique that identifies the behavior of the cyber-physical system, which deviates from normal behavior, then it can be determined that an anomaly is occurring. Different attack detection and mitigation techniques can be implemented for cyberattacks in the power electronics inverters and VSCs. Antivirus software (AVS) can be used to detect malicious code in SIs and embedded controllers. The AVS fits best for the low-level monolithic hardware systems such as the VSCs and power inverters. However, the AVS has drawbacks, like the attackers can create viruses that change the appearance of the code while keeping malicious functionality [68]. As the intelligent attackers can always create viruses that modify the appearances of the codes while maintaining the malicious functionality, so the implementation of the AVS requires large computation bandwidth and performance overhead. Moreover, AVS cannot always identify the malware that changes the execution algorithm in every iteration of propagation [69]. So, the attackers attempt to develop novel methods and algorithms, new appearances of the codes for evading the AVS, while the AVS seeks to bolster the defenses. Furthermore, the AVS cannot always detect the polymorphic and metamorphic malwares, which can change their execution format in each iteration.

Another potential mitigation method is the hardware malware detectors (HMDs), which can be utilized to detect and mitigate the adverse effect of cyber-attacks in the VSC of the SMES and power inverters [63, 70–75]. The HMD method utilizes hardware features and it is extremely challenging to compromise the operation of HMD as it operates on the actual hardware. To keep track of the low-level microarchitectural events, dedicated registers like HPC are used for the HMD. The HPC can determine the number of branch misses, CPU cycles, instruction execution iteration, etc. Furthermore, the HPC values collected during the cyberattack detection application execution can be utilized to train the ML methods too. The HPC can keep track of instructions inside the power inverter controller firmware, and the ML classifier can differentiate between the malicious modification and benign firmware. However, the application of HMDs directly into the microgrid is not always feasible due to the presence of legacy-embedded controllers.

A trust- and confidence-based resilient control approach for grid forming inverters against cyberattacks was discussed in References [76, 77], where each SI in the system monitors the information it receives from the neighbor, updates the local confidence factors and trust factors of all the neighboring SIs, and sends back to its neighbors. Data received at each inverter from the neighbor inverters are plugged into a distributed observer-based frequency update law, weighted by its trust factor and the confidence factor. The smaller confidence factor of the SI slows down the spread of corrupted data from the compromised SI to the neighbors. That eventually discards the information received from the compromised neighbors once the trust value drops below a predefined threshold. However, the trust- and confidence-based resilient control approach involves online calculation, which incurs additional computational burden. A cyber-resilient cooperative control for the bidirectional converter [78] based on the adaptive control theory can also be utilized for the SIs in different cyberattack scenarios.

Blockchain is a distributed database that maintains a continuous list of data records secured from tampering and revision [79]. Blockchain technology incorporating blockchain ledgers and smart contacts can be applied for energy trading, peer-to-peer transactions, demand side management, IoT security, privacy [80], etc. Blockchain technology for the SIs and VSCs can detect cyberattacks, improve connectivity, observability, and decentralized control toward a smarter environment.

For detecting and mitigating cyberattacks in the inverter-controlled resources of the power grid, different ML algorithms can be utilized as effective techniques. ML algorithms are artificial intelligence techniques that enable computers to learn about the system topology and improve without being explicitly programmed [81]. Different ML can be utilized including supervised learning, semi-supervised learning, unsupervised learning, and deep learning [81, 82].

The anomaly detection performance of different ML detection and mitigation techniques can be measured by determining different metrics. False positive (FP) refers to situations in which a benign operation is diagnosed as an anomaly operation, whereas false negative (FN) refers to situations in which an anomaly operation is misdiagnosed as a benign operation [83, 84]. Both are not desirable; however, a high FN rate means the anomaly detection algorithm is not able to detect malicious attacks, while high FP rate means the operator will have to deal with too many false alarms. On the other hand, true positive (TP) refers to situations in which a malicious operation is correctly diagnosed as an anomaly operation, whereas true negative (TN) refers to situations in which a benign operation is correctly classified as a benign operation. Based on the FP, FN, TP, and TN measurements, the Accuracy, Precision, Recall, and $F1$ score can be measured too as represented in equation 1. Accuracy is the proportion of the number of predictions correctly classified to the total number of predictions made. Accuracy is meaningful when the number of positive and negative measurement samples is equal. Precision represents the ratio of positive class classifications that is correct. Recall is the proportion of correct positive classifications to the total number of positive classifications. Recall is also termed as sensitivity and it gives detection model the capability to identify all attacks. $F1$ score is used to achieve an optimal tradeoff between the Precision and Recall and it is designed as the harmonic average of Precision and Recall.

$$
\left. \begin{array}{l}
\text{Accuracy} = \dfrac{\text{No. of } (\text{TP} + \text{TN})}{\text{No. of } (\text{FP} + \text{FN} + \text{TP} + \text{TN})} \\[3em]
\text{Precision} = \dfrac{\text{No. of TP}}{\text{No. of } (\text{FP} + \text{TP})} \\[3em]
Recall = \dfrac{\text{No. of TP}}{\text{No. of } (\text{FN} + \text{TP})}(1) \\[3em]
\text{F1 score} = \dfrac{2 * (\text{Recall} * \text{Precision})}{(\text{Recall} + \text{Precision})}
\end{array} \right\}
\qquad (8.2)
$$

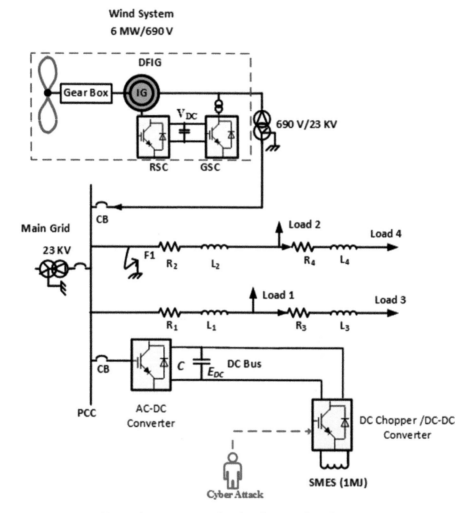

Figure 8.4   Microgrid with cyberattack in the SMES

8.7   Case study

The microgrid system model shown in Figure 8.4 has been used for simulation in this case study. This microgrid consists of double circuit distribution lines, which are delivering power to loads. The microgrid can operate in standalone mode and grid-connected mode while connected to 23 KV distribution substation. A 6 MW doubly fed induction generator (DFIG)-based wind farm is connected at the PCC of the microgrid through a step-up transformer. The microgrid line parameters and the wind generator parameters are presented in Table 7.2. The SMES unit is connected in the 23 KV PCC using power electronics interfaces like DC–DC converter and AC–DC converter for supporting the microgrid during voltage sag and voltage

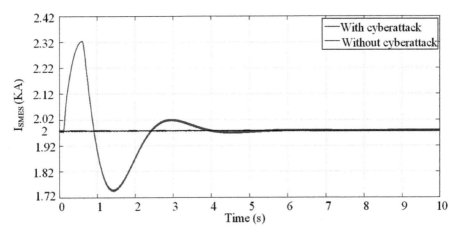

Figure 8.5 *SMES coil current response during normal operation and cyberattacks in the active power set points*

swell situations. The SMES is assumed to be fully charged at its maximum capacity of 1.0 MJ. The wind farm local distribution line and the substation double circuit distribution lines have the circuit breakers. The 6 MW wind farm is built from four 1.5 MVA wind turbines. The detailed design and modeling of RSC and GSC for the DFIG-based wind generator are available in References [85, 86].

In this case study, simulations have been performed by using the MATLAB/ Simulink software. Three attack situations are considered. *Attack A:* increase or decrease of the active power (P_{BUS}) set points of the primary controller of SMES by data injections. *Attack B:* Increase or decrease of the reactive power (Q_{BUS}) set points of the primary controller of SMES by data injections. Attack C: adding a DC bias to the sensed DC voltage (E_{dc}) measurements of the primary controller of the SMES. All these cyberattack situations happens during the standby mode of the SMES. For cyberattack simulations, consecutive samples are executed in each session of the MATLAB/Simulink environment. Data intrusions are designed as the time-controlled input signals from the library. For faster sampling and random generation of the attacks, the sampling rate is set to 20 μs.

The SMES coil current responses during the momentary cyberattack in the active power set points of the primary controller of SMES is presented in Figure 8.5. The cyberattack is considered to affect the controller only for 2 seconds and then the controller set points are settled to their original default settings. During normal operation, the SMES coil current is held constant at its rated value. However, during the cyberattacks, the coil current increases, giving a positive slope ($\frac{di}{dt}$) then decreases giving a negative slope ($\frac{di}{dt}$), and then settles to the desired rated value. This SMES coil current response indicates although rate of coil current change may be decided by the duty cycle of the DC–DC chopper during the normal operation but cyberattack attributes to the change in the coil current variations. Also, this transient

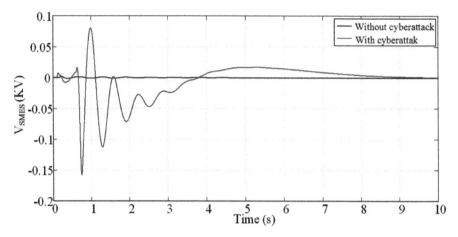

Figure 8.6   SMES coil voltage response during normal operation and cyberattacks in the reactive power set points

nature of the SMES coil current during the standby mode of operation is unsafe for the steady operation of the power grid.

The voltage across the SMES coil during normal operation is equal to zero during steady operation of the system and standby mode of the SMES as represented in Figure 8.6. During the momentary cyberattack in the reactive power set points, it is observed that the voltage across the SMES coil responds abruptly assuming charging and discharging mode of operation of the SMES even during the standby mode of operation. The magnitude of the voltage across the SMES coil is controlled by the duty cycle and DC link capacitor voltage of the SMES, which seems to have little effect as the set points are changed during cyberattacks.

The voltage across the DC link capacitor of the SMES is maintained at 8.7 KV during standby operation as represented in Figure 8.7. However, during cyberattacks in the reactive power set points, the DC link capacitor voltage changes abruptly indicating its ultimate effect on the SMES coil voltage and current. Furthermore, the abrupt nature of the DC link capacitor voltage hampers the proper operation of the AC DC converter and the voltage levels in the PCC.

During standby mode of operation, there is no transfer of energy between the SMES and the AC system as represented in Figure 8.8. However, as the cyberattack happens in the SMES primary controller active power set points, the energy is transferred from the AC system to the SMES until 0.9 seconds and then the SMES discharges the energy to the AC system until 2.5 seconds. Then again, the SMES coil energy increases but it then settles down to the 1 MJ energy storage. All these charging–discharging and settling down to expected level happened during the standby mode of operation in the presence of cyberattacks. It can be observed that the SMES coil current and voltage have a similar trend to the coil energy storage and their levels are correlated to each other even during the cyberattacks.

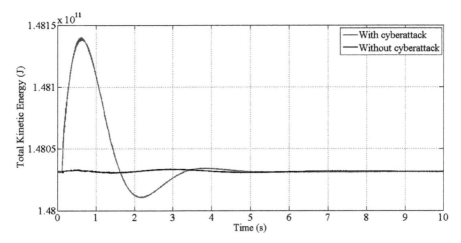

Figure 8.7   SMES DC link capacitor voltage response during normal operation and cyberattacks in reactive power set points

Figures 8.9.–8.10 represent the SMES DC link capacitor voltage and SMES coil voltage during addition of DC bias to the sensed DC voltage measurements of the primary controller. Thus, during cyberattacks, the response of the SMES becomes unsafe for the steady operation of the power grid. It may lead to unexpected operation of circuit breakers, increase the generator shaft speeds, and further accelerate the shutdown of generating sources in the grid.

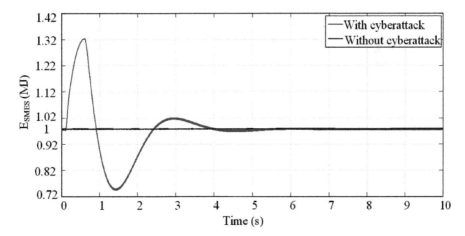

Figure 8.8   Stored energy of the SMES coil during normal operation and cyberattacks in active power set points

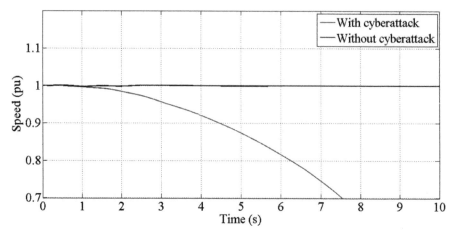

Figure 8.9   SMES DC link capacitor voltage response during normal operation and cyberattack in the sensed DC voltage of the primary controller

8.8   Challenges and future directions of cyber secure SI-enabled energy storage devices

The smart power electronics interfaces of the energy storage devices are vulnerable to cyberattacks and their control systems should be robust and reliable to keep the system's stability [87]. So, a strong coordination and interdependency are required among the inverter hardware, communication firmware, system components, and the actual legacy power grid itself [87]. Different challenges and future directions

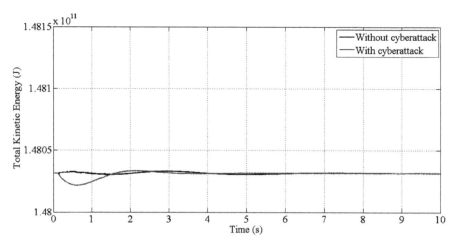

Figure 8.10   SMES coil voltage response during normal operation and cyberattack in the sensed DC voltage of the primary controller

of the cyber secure inverter-enabled energy storage device interconnection into the power grid have been discussed below.

8.8.1   Challenges

In inverter-enabled distribution systems, the DERs, energy storage devices, smart meters, etc. are interconnected with each other through a communication network, and it exposes several vulnerable nodes to the intruders. Furthermore, when the inverter-enabled resources are interconnected with the traditional synchronous generator-based legacy systems, it creates new opportunities for the intruders to inflict more damage in short span of time. Therefore, tight cyber-physical interactions and resiliency are required among all the constituent hardware, sub systems, cyber layers, etc. to thwart any cyberattacks.

One of the challenges associated with the application of the inverter interfaced ESS is its ability to properly distinguish between the cyberattacks and faults. Proper response of the ESS control logics during information uncertainties is also another challenge. Cyberattacks or impaired communications can corrupt the transmitted commands to the control logics of the ESS's SIs causing information uncertainties. These uncertainties can adversely affect the control layers of the power electronics interfaces and lead to ESS's inability to respond quickly to energy and power demands.

Interoperability of different power electronics inverters in a system is crucial for ensuring the cybersecurity. However, when several power electronics inverters of different vendors coordinate with each other, their syntactic compatibility reinforcements need to be monitored for the standardization of the practice and policies.

8.8.2   Future directions

As high penetration of power electronics-based inverters and ESSs in the power system poses challenges of low inertial in the system, it is vital to design control-level decisions based on the data generated in the system level. The ML methods are tools developed to make intelligent decisions based on the historical data. So, further research is needed to develop suitable methods based on the ML tools to further develop data-driven control logics and algorithms.

Hot patching is the ability to perform firmware update without causing any downtime in a device control unit. Hot patching can allow additional backup in the control layer when firmware vulnerabilities are being updated. More work needs to be done on the hot patching of the SI firmware.

8.9   Conclusion

This chapter discusses the cybersecurity issues in intelligent control-based SMES systems. Initially, different cybersecurity issues in the power electronics-enabled power grid have been discussed. In this section, different intrusion sources, how they can penetrate in different stages of the power grid, how intelligent control system can be affected, etc. have been discussed. Furthermore, how the SIs and VSCs in modern power grid can be a source of cyberattacks have been discussed. Next section discussed about the cybersecurity issues in the primary controllers of the SMES. How cyberattacks can take place in the SMES via the control system have been discussed explicitly. The impact of cyberattack on the SMES on the rest of the systems has been discussed too. It has been discussed that the cyberattacks in the SMES control system set points can initiate the unnecessary charging and discharging operation of the SMES. Also, cyberattacks on the SMES can hamper the operation of the rest of the system and may initiate the DoS operation. Furthermore, later in this chapter, different cyberattack detection and mitigation techniques have been discussed. An overview of different detection and mitigation techniques in protecting the SIs from cyberattacks has been discussed. A comprehensive case study has been presented in a microgrid environment for different cyberattacks situations in the primary control system of the SMES. Finally, challenges and future directions of cyber secure SI-enabled ESS devices have been discussed later in the chapter.

References

[1]   van Soest H.L., den Elzen M.G.J., van Vuuren D.P. 'Net-zero emission targets for major emitting countries consistent with the Paris agreement'. *Nature Communications*. 2021, vol. 12(1), pp. 1–9.

[2]   Aminifar F., Fotuhi-Firuzabad M., Safdarian A., Davoudi A., Shahidehpour M. 'Synchrophasor measurement technology in power systems: Panorama and state-of-the-art'. *IEEE Access: Practical Innovations, Open Solutions*. 2014, vol. 2, pp. 1607–28.

[3]   Liu Y., Zhan L., Zhang Y. 'Wide-area-measurement system development at the distribution level: an FNET/grideye example'. *IEEE Transactions on Power Delivery*. 2015, vol. 31(2), pp. 721–31.

[4]   Liu X., Shahidehpour M., Cao Y., Wu L., Wei W., Liu X. 'Microgrid risk analysis considering the impact of cyber attacks on solar pv and ESS control systems'. *IEEE Transactions on Smart Grid*. 2016, vol. 8, pp. 1330–39.

[5]   Zhang H., Meng W., Qi J., Wang X., Zheng W.X. 'Distributed load sharing under false data injection attack in an inverter-based microgrid'. *IEEE Transactions on Industrial Electronics*. 2018, vol. 66(2), pp. 1543–51.

[6]   Oxford Analytica. *Critical infrastructure sees rising cybersecurity risk*. Emerald Expert Briefings. Oxford Analytica; 2021.

[7]   *Executive order on improving the nation'S cybersecurity*. 2021. Available from https://www.whitehouse.gov/briefing-room/presidential-actions/2021/05/12/executive-order-on-improving-the-nations-cybersecurity/

[8]   Radoglou-Grammatikis P.I., Sarigiannidis P.G. 'Securing the smart grid: a comprehensive compilation of intrusion detection and prevention systems'. *IEEE Access: Practical Innovations, Open Solutions*. 2019, vol. 7, pp. 46595–620.

[9] Liu J., Xiao Y., Li S., Liang W., Chen C.L.P. 'Cyber security and privacy issues in smart grids'. *IEEE Communications Surveys & Tutorials*. 2012, vol. 14(4), pp. 981–97.

[10] Ghahremani E., Kamwa I. *IEEE Transactions on Energy Conversion*. 2011, vol. 26(4), pp. 1099–108.

[11] Taha A.F. *Secure estimation, control and optimization of uncertain*. West Lafayette: Purdue University; 2015.

[12] Kosut O., Jia L., Thomas R.J., Tong L. 'Malicious data attacks on the smart grid'. *IEEE Transactions on Smart Grid*. 2013, vol. 2(4), pp. 645–58.

[13] Valenzuela J., Wang J., Bissinger N. 'Real-time intrusion detection in power system operations'. *IEEE Transactions on Power Systems*. 2013, vol. 28(2), pp. 1052–62.

[14] Liu Y., Ning P., Reiter M.K. 'False data injection attacks against state estimation in electric power grids'. *ACM Transactions on Information and System Security*. 2013, vol. 14(1), pp. 1–33.

[15] Mousavian S., Valenzuela J., Wang J. 'Real-time data reassurance in electrical power systems based on artificial neural networks'. *Electric Power Systems Research*. 2013, vol. 96, pp. 285–95.

[16] Lee A. *Electric sector failure scenarios and impact analyses-draft*. National electric sector cybersecurity organization resource (NESCOR) technical Working group. 2015 Dec.

[17] Beg O.A., Johnson T.T., Davoudi A. 'Detection of false-data injection attacks in cyber-physical DC microgrids'. *IEEE Transactions on Industrial Informatics*. 2017, vol. 13(5), pp. 2693–703.

[18] Hao J., Kang E., Sun J., *et al.* 'An adaptive Markov strategy for defending smart grid false data injection from malicious attackers'. *IEEE Transactions on Smart Grid*. 2016, vol. 9(4), pp. 2398–408.

[19] Chen Y., Huang S., Liu F., Wang Z., Sun X. Evaluation of reinforcement learning-based false data injection attack to automatic voltage control. *IEEE Transactions on Smart Grid*. 2018, vol. 10(2), pp. 2158–69.

[20] Tan R., Nguyen H.H., Foo Eddy.Y.S, *et al.* Modeling and mitigating impact of false data injection attacks on automatic generation control. *IEEE Transactions on Information Forensics and Security*. 2017, vol. 12(7), pp. 1609–24.

[21] Tan R., Nguyen H.H., Foo E.Y.S, *et al.* 'Optimal false data injection attack against automatic generation control in power grids'. Presented at 2016 ACM/IEEE 7th International Conference on Cyber-Physical Systems (ICCPS); Vienna, Austria.

[22] Zhang J., Domínguez-García A.D. On the impact of measurement errors on power system automatic generation control. *IEEE Transactions on Smart Grid*. 2016, vol. 9(3), pp. 1859–68.

[23] Kumari R., Prabhakaran K.K., Chelliah T.R. Improved cybersecurity of power electronic converters used in hydropower plant. Presented at 2020 IEEE International Conference on Power Electronics, Drives and Energy Systems (PEDES); Jaipur, India.

[24]   Liu X., Ospina J., Konstantinou C. Deep reinforcement learning for cyberse-curity assessment of wind integrated power systems. *IEEE Access: Practical Innovations, Open Solutions*. 2020, vol. 8, pp. 208378–94.

[25]   Davari M., Nafisi H., Nasr M.-A., Blaabjerg F. A novel IGDT-based method to find the most susceptible points of cyberattack impacting operating costs of VSC-based microgrids. *IEEE Journal of Emerging and Selected Topics in Power Electronics*. 2020, vol. 9(3), pp. 3695–714.

[26]   Qi J., Hahn A., Lu X., Wang J., Liu C.-C. 'Cybersecurity for distributed energy resources and smart inverters'. *IET Cyber-Physical Systems*. 2016, vol. 1(1), pp. 28–39. Available from https://onlinelibrary.wiley.com/toc/23983396/1/1

[27]   Li F., Xie R., Yang B, *et al.* ' detection and identification of cyber and physi-cal attacks on distribution power grids with PVS: an online high-dimensional data-driven approach '. *IEEE Journal of Emerging and Selected Topics in Power Electronics*. 2022, vol. 10(1), pp. 1282–91.

[28]   Khan A., Hosseinzadehtaher M., Shadmand M.B., Saleem D., Abu-Rub H. Intrusion detection for cybersecurity of power electronics dominated grids: inverters PQ set-points manipulation. *2020 IEEE CyberPELS (CyberPELS)*; Miami, FL, USA, 2016. pp. 1–8.

[29]   Fard A.Y., Shadmand M.B., Mazumder S.K. Holistic multi-timescale attack resilient control framework for power electronics dominat-ed grid. *2020 Resilience Week (RWS)*; Salt Lake City, ID, USA, 2016. pp. 167–73. Available from https://ieeexplore.ieee.org/xpl/mostRecentIssue.jsp?punumber=9241210

[30]   Zanella A., Bui N., Castellani A., Vangelista L., Zorzi M. 'Internet of things for smart cities'. *IEEE Internet of Things Journal*. 2016, vol. 1(1), pp. 22–32.

[31]   Xia F., Yang L.T., Wang L., Vinel A. 'Internet of things'. *International Journal of Communication Systems*. 2012, vol. 25(9), pp. 1101–02. Available from http://doi.wiley.com/10.1002/dac.v25.9

[32]   Basso T., DeBlasio R. 'IEEE smart grid series of standards IEEE 2030 (in-teroperability) and IEEE 1547 (interconnection) status'. *National Renewable Energy Lab.(NREL)*; Golden, CO (United States), 2012.

[33]   Boemer J.C., Huque M., Seal B., Key T., Brooks D., Vartanian C. Status of revision of IEEE STD 1547 and 1547.1. *2017 IEEE Power & Energy Society General Meeting (PESGM)*; Chicago, IL, 2012. pp. 1–5.

[34]   Hadi A.A., Sinha U., Faika T., Kim T., Zeng J., Ryu M.-H. 'Internet of things (iot) -enabled solar micro inverter using blockchain technology'. *2019 IEEE Industry Applications Society Annual Meeting*; Baltimore, MD, USA, 2012. pp. 1–5. Available from https://ieeexplore.ieee.org/xpl/mostRecentIssue.jsp?punumber=8894477

[35]   McLaughlin S., Konstantinou C., Wang X. The cybersecurity landscape in industrial control systems. *Proceedings of the IEEE*. 2012, vol. 104(5), pp. 1039–57.

[36]   Ma R., Cheng P., Zhang Z., Liu W., Wang Q., Wei Q. Stealthy attack against redundant controller architecture of industrial cyber-physical system. *IEEE Internet of Things Journal*. 2019, vol. 6(6), pp. 9783–93.

[37] Tan S., Guerrero J.M., Xie P., Han R., Vasquez J.C. Brief survey on attack detection methods for cyber-physical systems. *IEEE Systems Journal*. 2020, vol. 14(4), pp. 5329–39.

[38] Liu J., Singh R., Pal B.C. Distribution system state estimation with high penetration of demand response enabled loads. *IEEE Transactions on Power Systems*. 2021, vol. 36(4), pp. 3093–104.

[39] Manandhar K., Cao X., Hu F., Liu Y. Detection of faults and attacks including false data injection attack in smart grid using Kalman filter. *IEEE Transactions on Control of Network Systems*. 2014, vol. 1(4), pp. 370–79.

[40] Chen J., Patton R.J. Robust model-based fault diagnosis for dynamic systems. Vol. 3. Springer Science & Business Media; 2012.

[41] Zhang J., Yao H., Rizzoni G. Fault diagnosis for electric drive systems of electrified vehicles based on structural analysis. *IEEE Transactions on Vehicular Technology*. 2016, vol. 66(2), pp. 1027–39.

[42] Giani A., Elasser A. 'A review of cyber-physical security for photovoltaic systems'.2021.

[43] Ren X., Yan J., Mo Y. Binary hypothesis testing with Byzantine sensors: fundamental tradeoff between security and efficiency. *IEEE Transactions on Signal Processing*. 2018, vol. 66(6), pp. 1454–68.

[44] Kailkhura B., Han Y.S., Brahma S., Varshney P.K. Asymptotic analysis of distributed Bayesian detection with Byzantine data. *IEEE Signal Processing Letters*. 2014, vol. 22(5), pp. 608–12.

[45] Wei F., Wan Z., He H. 'Cyber-attack recovery strategy for smart grid based on deep reinforcement learning'. *IEEE Transactions on Smart Grid*. 2019, vol. 11(3), pp. 2476–86.

[46] Wang H., Ruan J., Wang G, *et al.* Deep learning-based interval state estimation of Ac smart grids against sparse cyber attacks. *IEEE Transactions on Industrial Informatics*. 2018, vol. 14(11), pp. 4766–78.

[47] Li F., Shi Y., Shinde A., Ye J., Song W. Enhanced cyber-physical security in Internet of things through energy auditing. *IEEE Internet of Things Journal*. 2019, vol. 6(3), pp. 5224–31.

[48] Karpilow A., Cherkaoui R., D'Arco S., Duong T.D. Detection of bad PMU data using machine learning techniques. Presented at 2020 IEEE Power & Energy Society Innovative Smart Grid Technologies Conference (ISGT); Washington, DC, USA. Available from https://ieeexplore.ieee.org/xpl/mostRecentIssue.jsp?punumber=9078689

[49] Lore K.G., Shila D.M., Ren L. Detecting data integrity attacks on correlated solar farms using multi-layer data driven algorithm. Presented at 2018 IEEE Conference on Communications and Network Security (CNS); Beijing. Available from https://ieeexplore.ieee.org/xpl/mostRecentIssue.jsp?punumber=8410986

[50] Karimipour H., Dehghantanha A., Parizi R.M., Choo K.-K.R., Leung H. 'A deep and scalable unsupervised machine learning system for cyber-attack detection in large-scale smart grids'. *IEEE Access: Practical Innovations, Open Solutions*. 2019, vol. 7, pp. 80778–88.

[51] Esmalifalak M., Liu L., Nguyen N., Zheng R., Han Z. 'Detecting stealthy false data injection using machine learning in smart grid'. *IEEE Systems Journal*. 2014, vol. 11(3), pp. 1644–52.

[52] Yu Z.-H., Chin W.-L. 'Blind false data injection attack using PCA approximation method in smart grid'. *IEEE Transactions on Smart Grid*. 2015, vol. 6(3), pp. 1219–26.

[53] Sakhnini J., Karimipour H., Dehghantanha A. Smart grid cyber attacks detection using supervised learning and heuristic feature selection. Presented at 2019 IEEE 7th International Conference on Smart Energy Grid Engineering (SEGE); Oshawa, ON, Canada.

[54] Li Q., Li F., Zhang J., Ye J., Song W., Mantooth A. Data-Driven cyberattack detection for photovoltaic (PV) systems through analyzing micro-pmu data. Presented at 2020 IEEE Energy Conversion Congress and Exposition (ECCE); Detroit, MI, USA. Available from https://ieeexplore.ieee.org/xpl/mostRecentIssue.jsp?punumber=9235288

[55] Hupp W., Hasandka A., de Carvalho R.S., Saleem D. 'Module-OT: a hardware security module for operational technology'. Presented at 2020 IEEE Texas Power and Energy Conference (TPEC); College Station, TX, USA.

[56] Konstantinou C., Maniatakos M. Impact of firmware modification attacks on power systems field devices. *2015 IEEE International Conference on Smart Grid Communications (SmartGridComm)*; Miami, FL, USA, 2020. pp. 283–88.

[57] Johnson J., Onunkwo I., Cordeiro P., Wright B.J., Jacobs N., Lai C. 'Assessing Der network cybersecurity defences in a power-communication co-simulation environment'. *IET Cyber-Physical Systems*. 2020, vol. 5(3), pp. 274–82. Available from https://onlinelibrary.wiley.com/toc/23983396/5/3

[58] Saleem D., Sundararajan A., Sanghvi A., Rivera J., Sarwat A.I., Kroposki B. 'A multidimensional holistic framework for the security of distributed energy and control systems'. *IEEE Systems Journal*. 2019, vol. 14(1), pp. 17–27.

[59] Jones C.B., Chavez A.R., Darbali-Zamora R., Hossain-McKenzie S. Implementation of intrusion detection methods for distributed photovoltaic inverters at the grid-edge. *2020 IEEE Power & Energy Society Innovative Smart Grid Technologies Conference (ISGT)*; Washington, DC, USA, 2020. pp. 1–5. Available from https://ieeexplore.ieee.org/xpl/mostRecentIssue.jsp?punumber=9078689

[60] White J.S., Fitzsimmons T., Matthews J.N., Ternovskiy I.V., Chin P. Quantitative analysis of intrusion detection systems: snort and suricata. *SPIE Defense, Security, and Sensing*; Baltimore, Maryland, USA, 2020.

[61] Ramos-Ruiz J., Kim J., Ko W.-H. An active detection scheme for cyber attacks on grid-tied PV systems. *2020 IEEE CyberPELS (CyberPELS)*; Miami, FL, USA, 2021. pp. 1–6.

[62] Jhala K., Pradhan P., Natarajan B. 'Perturbation-based diagnosis of false data injection attack using distributed energy resources'. *IEEE Transactions on Smart Grid*. 2020, vol. 12(2), pp. 1589–601.

[63] Kuruvila A.P., Zografopoulos I., Basu K., Konstantinou C. 'Hardware-assisted detection of firmware attacks in inverter-based cyberphysical microgrids'. *International Journal of Electrical Power & Energy Systems*. 2021, vol. 132, p. 107150.

[64] Seshadri S.S., Rodriguez D., Subedi M., *et al.* 'IOTCOP: a blockchain-based monitoring framework for detection and isolation of malicious devices in internet-of-things systems'. *IEEE Internet of Things Journal*. 2020, vol. 8(5), pp. 3346–59.

[65] Mylrea M., Gourisetti S.N.G. 'Blockchain for supply chain cybersecurity, optimization and compliance'. *2018 Resilience Week (RWS)*; Denver, CO, 2021. pp. 70–76.

[66] Choi J., Narayanasamy D., Ahn B., Ahmad S., Zeng J., Kim T. A real-time Hardware-in-the-Loop (Hil) cybersecurity testbed for power electronics devices and systems in cyber-physical environments. *2021 IEEE 12th International Symposium on Power Electronics for Distributed Generation Systems (PEDG)*; Chicago, IL, USA, 2013. pp. 1–5.

[67] Kim T., Ochoa J., Faika T, *et al.* 'An overview of cyber-physical security of battery management systems and adoption of blockchain technology'. *IEEE Journal of Emerging and Selected Topics in Power Electronics*. 2020, vol. 10(1), pp. 1270–81.

[68] Murad K., Shirazi S.N.-U.-H., Zikria Y.B., Ikram N. ' evading virus detection using code obfuscation '. Presented at in International Conference on Future Generation Information Technology; 2010: Springer, Berlin, Heidelberg,

[69] Demme J., Maycock M., Schmitz J, *et al.* On the feasibility of online malware detection with performance counters. *ACM SIGARCH Computer Architecture News*. 2013, vol. 41(3), pp. 559–70.

[70] Ozsoy M., Khasawneh K.N., Donovick C., Gorelik I., Abu-Ghazaleh N., Ponomarev D. 'Hardware-based malware detection using low-level architectural features'. *IEEE Transactions on Computers*. 2013, vol. 65(11), pp. 3332–44.

[71] Malone C., Zahran M., Karri R. Are hardware performance counters a cost effective way for integrity checking of programs. *The Sixth ACM Workshop*; Chicago, Illinois, USA, New York, New York, USA, 2017. pp. 71–76. Available from http://dl.acm.org/citation.cfm?doid=2046582

[72] Wang X., Konstantinou C., Maniatakos M., Karri R. 'Confirm: detecting firmware modifications in embedded systems using hardware performance counters'. *2015 IEEE/ACM International Conference on Computer-Aided Design (ICCAD)*; Austin, TX, USA, 2017. pp. 544–51.

[73] Singh B., Evtyushkin D., Elwell J., Riley R., Cervesato I. 'On the detection of kernel-level rootkits using hardware performance counters'. *ASIA CCS '17*; Abu Dhabi United Arab Emirates, New York, NY, USA, 2017. pp. 483–93. Available from https://dl.acm.org/doi/proceedings/10.1145/3052973

[74] Sayadi H., Makrani H.M., Pudukotai Dinakarrao S.M. '2SMaRT: a two-stage machine learning-based approach for run-time specialized hardware-assisted

malware detection'. *2019 Design, Automation & Test in Europe Conference & Exhibition (DATE)*; Florence, Italy, 2017. pp. 728–33.

[75] Wang X., Konstantinou C., Maniatakos M. 'Malicious firmware detection with hardware performance counters'. *IEEE Transactions on Multi-Scale Computing Systems*. 2017, vol. 2(3), pp. 160–73.

[76] Abhinav S., Modares H., Lewis F.L., Ferrese F., Davoudi A. 'Synchrony in networked microgrids under attacks'. *IEEE Transactions on Smart Grid*. 2017, vol. 9(6), pp. 6731–41.

[77] Sahoo S., Dragicevic T., Yang Y., Blaabjerg F. 'Adaptive resilient operation of cooperative grid-forming converters under cyber attacks'. Presented at 2020 IEEE CyberPELS (CyberPELS); Miami, FL, USA. Available from https://ieeexplore.ieee.org/xpl/mostRecentIssue.jsp?punumber=9311528

[78] Wang Y., Mondal S., Deng C., Satpathi K., Xu Y., Dasgupta S. 'Cyber-resilient cooperative control of bidirectional interlinking converters in networked AC/DC microgrids'. *IEEE Transactions on Industrial Electronics*. 2020, vol. 68(10), pp. 9707–18.

[79] Prusty N. *Building blockchain projects*. Birmingham, UK: Packt Publishing Ltd; 2017.

[80] Christidis K., Devetsikiotis M. 'Blockchains and smart contracts for the Internet of things'. *IEEE Access: Practical Innovations, Open Solutions*. 2016, vol. 4, pp. 2292–303.

[81] Musleh A.S., Chen G., Dong Z.Y. 'A survey on the detection algorithms for false data injection attacks in smart grids'. *IEEE Transactions on Smart Grid*. 2019, vol. 11(3), pp. 2218–34.

[82] Ozay M., Esnaola I., Yarman Vural F.T., Kulkarni S.R., Poor H.V. 'Machine learning methods for attack detection in the smart grid'. *IEEE Transactions on Neural Networks and Learning Systems*. 2016, vol. 27(8), pp. 1773–86.

[83] Sridhar S., Govindarasu M. 'Model-based attack detection and mitigation for automatic generation control'. *IEEE Transactions on Smart Grid*. 2014, vol. 5(2), pp. 580–91.

[84] Chen C., Zhang K., Yuan K., Zhu L., Qian M. 'Novel detection scheme design considering cyber attacks on load frequency control'. *IEEE Transactions on Industrial Informatics*. 2017, vol. 14(5), pp. 1932–41.

[85] Boukhezzar B., Siguerdidjane H. 'Nonlinear control of a variable-speed wind turbine using a two-mass model'. *IEEE Transactions on Energy Conversion*. 2011, vol. 26(1), pp. 149–62.

[86] Okedu K.E., Muyeen S.M., Takahashi R., Tamura J. 'Wind farms fault ride through using DFIG with new protection scheme'. *IEEE Transactions on Sustainable Energy*. 2012, vol. 3(2), pp. 242–54.

[87] Mazumder S.K., Kulkarni A., Sahoo S, *et al.* 'A review of current research trends in power-electronic innovations in cyber–physical systems'. *IEEE Journal of Emerging and Selected Topics in Power Electronics*. 2021, vol. 9(5), pp. 5146–63.

Chapter 9

Outlook

P. Mukherjee and Prof. Rao V. V.[1]

9.1 Introduction

This book describes the superconducting magnetic energy storage (SMES) system, its operation and control to contribute in various multifunctional performances in power grid. It has been observed that SMES has the ability to overcome a number of issues in conventional power grid as well as nonconventional microgrids. Till date, the realization of SMES systems is limited to its development in solenoid and toroid geometry of low temperature superconducting and high temperature superconducting materials. These SMES are capable of storing huge amount of energy within its superconducting coil. To realize the feasibility of charging and discharging of SMES energy according to its application in power grid substantial research is in progress. An SMES unit consists of superconducting coil in cryogenic and electromagnetic environment, power converter circuits and control techniques. Therefore, the feasibility of SMES depends upon the research and development of these essentials. SMES has advantages of high-energy storage efficiency and large discharge within a short period of time. This makes SMES widely used in the power distribution systems with high penetration of renewable energy resources to improve the dynamic performances of the system. In **Chapter 2**, the basics of SMES are introduced. Different energy storage technologies as well as the operating principle and the main components are presented. Different geometrical configurations and the electromagnetic forces and stresses are explained. The theorems and factors that involve the variation of shapes of coil with stored energy are presented. SMES system is compared with other technologies to find out the advantages and drawbacks. A review on dynamic model and applications of SMES on power grid is given. An estimation of SMES system is presented based on material, configuration and application. **Chapter 3** explains the integration of SMES systems in the AC power grids for a specific application, which needs dynamical compensation of active and/or reactive power. The SMES integration needs converter topologies such as voltage source converter (VSC) and

[1]University of Engineering and Management, India, Kharagpur
[1]Indian institute of Technology, Kharagpur

current source converter (CSC). These converters are controlled with different control techniques for the VSC and CSC. The different control methodologies for VSC and CSC are used to mitigate the variation in voltage and power for grid-connected systems using SMES. For VSC-based SMES, adaptable proportional integral (PI) controller, fuzzy logic control, linear control by state feedback and model predictive controller (MPC) are used. For CSC-based control, decoupled state feedback control, linear-quadratic regulator control, power control theory and nonlinear feedback are used. Power quality disturbances occur due to the inherent nonlinear characteristics and stochastic nature of power grids. Power quality issues occur at various levels of the grid-like power plants, transmission system, substations, service equipment and building wiring. Low-quality power can harm and destroy the power grid equipment. Flexible alternating current transmission system (FACTS) devices are the conventional method to mitigate the power quality issues. In **Chapter 4,** the transient stability issues in power grid are discussed. The conventional way to improve transient stability is explained. The ability of SMES in enhancing the transient stability of power grids is studied. The intermittent power outputs of wind energy sources are anticipated during the stability study. The basic model of SMES encompassing the superconducting coil and the converter is described. The detailed mathematical modelling of various SMES control algorithms, such as derivative control, discrete control, neural network and intelligent fuzzy logic, is produced. Case studies on time-domain simulations are performed on the Western System Coordinating Council and the 70 bus, 18 machine system. **Chapter 5** introduces the load frequency control issues in interconnected microgrid systems. First, the frequency deviation in the microgrid system is formulated, and the SMES is modelled accordingly. Then, the various utilized devices and control methods in the interconnected microgrid for frequency regulation are introduced. Frequency and tie-line power oscillations are minimized in a microgrid connected with conventional generation systems and high voltage direct current systems using SMES-based FACTS devices. On the other hand, the frequency and tie-line power oscillations are minimized in renewable energy source-based interconnected microgrids by using photovoltaic farms, wind plants, electric vehicles and energy storage devices. The control methods of SMES considered for load frequency control study are integer order controllers, fractional order controllers, fuzzy logic controllers, data-driven load frequency control (LFC) methods and model predictive controllers. **Chapter 6** describes the design configurations and operation of SMES in combination with other ancillary devices like FACTS, battery energy storage system (BESS), fault current limiter (FCL), etc. These configurations are supposed to improve the flexibility and give superior performance than that of the SMES alone. In this configuration, the advantages of the individual devices can be achieved. FACTS devices with SMES are developed for improving the transient stability margin and damping the power system oscillations. The high-power density of SMES combined with high-energy density of BESSs improves the flexibility during dynamic stability study. SMES and FCL can improve the dynamic stability in power grid by reducing the higher transient current. The operation principles of SMES and different auxiliary devices have been discussed with their control systems. In **Chapter 7,** different controllers are applied to SMES for utmost enhancement of fault ride through (FRT) capability of doubly fed induction generator

(DFIG)-based wind generator system. Different artificial intelligent controllers (AICs) are designed, modelled and implemented for SMES. A comparative study has been carried out to suggest a superior AIC for SMES to improve the FRT capability of DFIG. Finally, in **Chapter 8,** cybersecurity issues in intelligent control-based SMES systems are discussed. Communication is required for the real-time measurement of power system variables for proper operation and maintenance. The cyberattack intrudes unnecessary data into the communication lines. This affects the performance of power system and disturbs the stability of the system. Furthermore, the cyberattack also creates problem in control signals of SMES, which leads to the malfunctioning of SMES system. This gives additional risks to the stability of power system. The smart VSCs with support of Internet of things and intelligent controller are solutions to these issues.

9.2   Summary of findings

SMES is a high-power device, which has a short discharging time (several seconds), high density of stored energy and high-power density. The current density is 100 times larger than copper conductors. The number of charge–discharge cycles is very high. Controlled output active and reactive power from SMES can enhance the stability and quality of power. It is a static device, therefore, has high-energy conversion efficiency.

For same amount of energy, toroid needs fewer conductors than solenoid. Losses per cycle per unit length of conductor are slightly higher in case of the toroid. This is due to the higher transport current of toroid compared with solenoid. The solenoid has a high magnetic field, which reduces the transport current and thus reduces the losses. From a materials point of view, Yttrium Barium Copper Oxide-based SMES coil not only has the advantages of smaller magnet size, less total conductor length and smaller AC losses but also has high magnetic flux and very high mechanical stresses than Bismuth Srontium Calcium Copper Oxide, MgB_2, etc. For a small-capacity (kJ) superconducting magnet, solenoid coil is preferred. For a medium or large SMES system, a toroid configuration is a better option. Bigger radius magnet has low magnetic fields, high current and more storage capacity. Bigger radius magnet requires longer conductor length and high current increases AC losses. Therefore, it is required to optimize the size of the conductor with respect to energy capacity and coil stability.

The power quality of power grid consisting of the two machines, nine bus IEEE power system and DFIG is studied. The automatic voltage regulator and governor control models for alternators are considered for this study. The power quality is disturbed due to the occurrence of fault at three different locations. SMES can enhance the power quality of the grid.

Due to its high-energy storage capability, SMES with inverter filter and controller can act as a virtual synchronous generator and take part to increase transient stability in terms of rotor angle oscillations. The transient stability is heuristically improved in the case of fuzzy-controlled SMES. SMES-based VSG improves frequency deviation considerably, improves the voltage profile in multimachine system and minimizes voltage overshoot. Discrete predictive controlled distributed SMES system reduces the size and installation cost of energy storage. Distributed energy storage system topology provides constant DC link voltage status. Reactive power oscillations are considerably lower after applying SMES, which improve the voltage profile of the system. SMES gives abundant support to the primary frequency control encounters after load fluctuation. The transient stability support, i.e., power overshoot and power smoothing, is effectively much higher using SMES with parametric fuzzy predictive controller strategy, since the rotor angles face minimal excursions.

The performance of LFC of a system, consisting of two areas, is improved with SMES. One area includes thermal power plant, wind generator, SMES and local loads. The other one contains hydraulic power plant, photovoltaics generators, SMES and local load. Different control structures have been selected for LFC such as proportional integral derivative (PID), fractional order PID (FOPID), tilt integral derivative (TID) and hybrid controller. All the controller parameters have been optimized using the manta ray foraging optimisation algorithm to minimize the objective function of the hybrid controller. The SMES maintains system frequency with decreased overshoot and settling time compared with the other controllers. The PID, FOPID and TID controllers suffer from frequency oscillations around the steady state value, and sometimes a higher overshoot appears during load changes. The hybrid controller of SMES also has the ability to keep system frequency at the original values compared with the other controllers.

A comparative study has been carried out between AICs and PI controllers to maximize the effectivity of SMES for enhancing the low voltage ride through (LVRT) capability of DFIG system. The implementation of controllers is thoroughly discussed. It can be established that the performance of DFIG during fault can be considerably improved by the proposed AICs of SMES. Moreover, the performance of the proposed adaptive neuro fuzzy controller (ANFC) controlled SMES is much better than that of PI, artificial neural network (ANN) and fuzzy logic controller (FLC)-controlled SMES in order to improve the dynamic stability of wind generator. The technical significance of this study is explained below.

The AICs are effective, simple and easy to implement to SMES converter than PI controller. The functionality of AICs is more understandable than PI controller. SMES is minimising voltage deviation and power deviation during the fault. AIC controller gives better result than PI controller. Among different AICs, ANFC gives superior control for SMES. In stored energy discharge and power exchange perspective, FLC is better than ANN even though performances in voltage and power deviations are almost same. However, adaptive neuro fuzzy interface system requires more power and energy than FLC but less than ANN to eliminate voltage deviation for LVRT. This proves that when FLC is trained to adapt the system response through ANN, it improves the effectiveness of controller.

During cyberattack, unnecessary charging and discharging may take place as the active and/or reactive power set point of SMES controller regulates due to data injections. Furthermore, a DC bias may needlessly add to the controller sensing the DC link voltage. Therefore, increase or decrease in the coil current occurs without any reason. SMES coil current should be controlled, and its variations may be decided by the duty cycle of DC-DC chopper. This control is affected by cyberattack, and uncontrolled coil current variations occur. All these actions happen during the standby conditions of SMES. These malfunctions of SMES lead to instability in power grid. It may cause unpredicted operation of circuit breakers and increase in generator shaft speeds to that extent, which needs shut down of generating sources in the grid.

9.3 Open challenges and unsolved problems

The power electronics converter of SMES, interfacing the coil with the AC line, needs to control according to power system requirements and superconducting coil protection. At numerous conditions, to maintain power quality and reliability, SMES needs to release very high power with high di/dt rate. On the other hand, reduced di/dt is preferred from the AC loss point of view to prevent the quenching of the superconducting coil. SMES converter with low di/dt is required for voltage sag and swell, primary frequency stability and load frequency control. Other than these, SMES needs more research to overcome the following issues:

- SMES discharges very high currents in a very short period of time. The connected converters circuit has to deal with this high current.
- Multiple solenoid type shows very good characteristics on stray field in comparison with solenoid type, but it has very poor energy density. A drawback of solenoid configuration is the stray field, which is a leakage magnetic field and presents a threat to environment and human health.
- High powers require large currents and excellent electric isolation for high voltages.
- At normal conductive state the resistivity of superconductor is high at temperatures higher than the critical value.
- The number of charge–discharge cycles is very high being mainly limited by the mechanical fatigue of the support structure.
- The requirement of high critical temperature (Tc) and transportation of high electrical currents in a high magnetic field are yet to be solved.
- Estimation of the cost of a superconducting coil system and the cooling system needed to keep it cold is often difficult to get budgetary.

9.4 Conclusion

This book reports detailed design of various configuration aspects, analytical studies on various applications and controllers of SMES. Implementation of the advanced controller for SMES, the operational effect of SMES with advanced power electronic converters and the locational significance of SMES in a power grid or microgrid are very important concerns. However, SMES technology needs more attention and research towards commercialisation as it is an amalgamation of electrical, vacuum, mechanical, materialistic and cryogenic technologies. Therefore, continuation and rigorous research are required to overcome the issues of SMES in every aspect. Due to the high cost of superconducting conductors, it is very expensive to develop SMES coil, which is the main component of SMES system. Low-cost superconducting material with transport current at atmospheric temperature is the primary requirement of this system to do more experimental research, which leads this to commercialisation and solves all the issues of power grid as discussed in this book.

Index